Soil

Håkan Wallander

Soil

Reflections on the Basis of our Existence

Springer

Håkan Wallander
Department of Biology,
Microbial Ecology Group
Lund University
Lund
Sweden

Translated by Greg McIvor and Nicole Tyszkiewicz.

ISBN 978-3-319-35808-6 ISBN 978-3-319-08458-9 (eBook)
DOI 10.1007/978-3-319-08458-9

Springer Cham Heidelberg New York Dordrecht London

Softcover reprint of the hardcover 1st edition 2014

Printed on acid-free paper

Springer is part of Springer Science+Business Media (www.springer.com)

Foreword

I have always enjoyed getting my fingers dirty, not just through gardening but also when out walking in the woods collecting samples for my research. I love rummaging among leaves and inhaling the aroma of mushrooms. Sometimes I will lift up the earth next to a large boulder to examine the tree roots. Some are yellow, others pink or black, all extending outwards in countless shapes and forms. The fungal mycelia that grow out of roots can penetrate the tiniest cavities in the soil and help the tree absorb nutrients, in return gaining access to the sugar the tree produces during photosynthesis. I have studied these fungi for 22 years, the first 7 at the Swedish University of Agricultural Sciences in Uppsala and the last 15 at Lund University.

As a scientist I have visited many exciting places: South America's rainforests, African desert, North American glaciers and the Scandinavian tundra. My travels have given me many different experiences and brought me into contact with both natural beauty and numerous interesting and knowledgeable people. Countless questions have been raised, and many are the nights I have lain awake and thought about measurements I would like to take and experiments I would like to do.

Some scientists are experts in specific soil organisms or the chemical or physical properties in the soil that relate to the life forms that grow in it. Others are more generalist in their attempts to understand how different soil components interact and how they affect critical processes like decomposition and nutrient circulation.

Some of these scientists are mentioned in this book, either in the narrative or in the references.

The book begins with a chapter on soil's importance for human life and how critical it is to preserve its properties. The chapters that follow look at how soil is formed and how soil quality affects what grows in it. The "Life in the Soil" chapter presents the life forms that inhabit the soil and the following chapter looks at how their activities enable nutrients to circulate in the soil so the wheels of life can turn. "Biological Warfare" and "A Matter of Taste" offer practical examples of how soil organisms can help us and how the soil affects the taste and quality of our food. The final chapter looks towards the future and examines how the way we manage the soil may have a profound impact on climate. This chapter may be a little harder to read, but it is a complex subject and one containing some of the toughest nuts for tomorrow's climate scientists to crack.

The pages intersperse empirical knowledge with stories from my travels and everyday life. Some of this knowledge comes from my own scientific experience, in other cases from the scientific literature or colleagues. My ambition has not been to cover all the different explanations and hypotheses for the different phenomena I describe, but rather to stimulate interest and curiosity in a subject of fundamental importance for human existence: the soil in which we grow our food and that is the basis for all land-based ecosystems. As textile artist Anna-Lisa Menander once aptly said, "You are not dirty when you have soil on your hands, you are holding a piece of eternity."

Contents

Our Indispensable Soil

As the bee collects nectar
and departs without injuring the flower
or its colour or scent,
so let a sage dwell in the community.

Buddha

Abstract This chapter describes the significance of soil for humanity. Concepts such as sustainability and nutrient balance are discussed. Most plants live in symbiosis with mycorrhizal fungi which spread their finely branched mycelium in the ground and facilitate the uptake of nutrients but the fungal hyphae also help to bind soil particles together into stable aggregates, providing a loose soil structure which is beneficial to the vegetation. Erosion, desertification and salinisation are global problems that arise when we use the land in the wrong way. My own experiences from research in Nepal and Tunisia serve here as current cases of such problems and how they can be resolved, along with historical examples of American farmers forced to abandon their farms and try their luck in California. Lastly, the chapter discusses organic farming from a sustainability perspective.

When attending a meeting at Bristol University, I was treated to a rare moment of popular science in a usually strictly academic environment. We were listening to a PowerPoint presentation full of graphs and tables describing the state of Europe's soil when a grey-bearded man wearing a woollen waistcoat approached the lectern. Squinting at the audience, he held up an apple grown organically in his garden and said it represented the Earth. He cut it in half with a knife and told us one half represented the oceans, with no soil for cultivation, and handed this piece to a colleague sitting nearby, who proceeded to take a bite. He sliced the remaining half in two and set one piece aside, saying it represented mountainous and polar terrain too cold or inhospitable for agriculture. After discarding one-eighth of the apple as representing urban areas, roads and polluted areas where crop growing was also impossible, he was left with a small segment that he carefully peeled. "This is the arable soil left for us to live on," he said, holding the thin sliver up. "We need to take care of it."

H. Wallander, *Soil*, DOI 10.1007/978-3-319-08458-9_1

What is Sustainable Agriculture?

All great civilisations have built their prosperity on a long tradition of farming fertile soil. Along the River Nile, rich volcanic silt washed by rainwater from the mountains of Abyssinia enabled farmers to cultivate the surrounding flood plains for thousands of years of agriculture without loss of soil fertility. The silt deposits came from once-forested slopes cleared by poor Abyssinian shepherds to create pasture for their animals. By destroying the forests, the shepherds exposed the earth to erosion and inadvertently ensured the viability of agriculture in Egypt. To this day, millions of tonnes of rich sediment accumulate on the Nile's flood plains and deltas every year.

Sustainable agriculture means ensuring that nutrients removed from the land during harvesting are replenished in equal measure. This can be through the addition of fertiliser or by flooding, as in the case of the Nile. Sustainable farming also requires that soil humus is not depleted because it is important for maintaining the soil's structure and ability to retain water. Humus decomposes at a higher rate when soil is cultivated but will increase if compost is added or plants with large underground root systems are grown. If no plant matter grows on the soil, it becomes vulnerable to erosion by wind and rain.

Inexorable soil depletion from agriculture. Soil fertility declines if the nutrients lost at harvest time are not replenished. Since 1843, scientists at Rothamsted research park in the UK have been studying how the presence or lack of nutrients affects crop yields. The bright yellow rows have had no nitrogen added to them, resulting in a yield of only 10 % of that produced in the fully fertilised rows

The Constituents of Soil

If you destroy the structure of the soil, with its delicate networks of pores through which air and water pass, you destroy the soil itself. The best way to preserve the soil's structure is to let plants grow in it. Plants protect the soil against raindrops, while dead plant matter in the soil supports fungi and bacteria that help keep the soil particles together and maintain a stable structure.

Soil is made up of mineral particles, organic substances, air, water and living organisms. Pick up a handful of soil and you will notice that it consists of lumps of varying sizes. These are known as aggregates and can be as small as a tenth of a millimetre across. Aggregates bond to create larger lumps that together give the soil its porosity. Without them, the soil would be dense and impermeable. Aggregates are formed by mineral particles that adhere to organic matter via slime secreted by fungi and bacteria. The excrement of earthworms is beneficial to soil because the worms' stomachs are filled with slime-creating bacteria. The aggregates are held together by thread-like fungi hyphae, which are also coated in water-repellent substances that prevent the soil from collapsing during rainfall.

The soil's network of pores is essential to enable air to reach plant roots and oxygen-reliant organisms and for water to penetrate the earth. Digging by terrestrial animals helps porosity by creating hollows in the ground, but pores are most concentrated around plant roots, where biological activity is greatest. Roots secrete carbohydrates and amino acids on which a rich microflora of bacteria and fungi depends, in turn providing food for amoebae and other small organisms that live in the soil.

Fungi that exist in a state of symbiosis with plant roots have been my specialist area of scientific study for 20 years. They are known as mycorrhizal fungi, a word that comes from the Greek *myko* (fungus) and *rhiza* (root). Ultra-thin fungal hyphae are adept at penetrating all the nooks and crannies of the soil in their quest for nutrients. They sprout from the tiniest plant roots and extend in complicated networks that also play their part in stabilising the soil structure. I like to ask my students to analyse a handful of soil from a pine forest under the microscope so they see just how ubiquitous hyphae are: they are everywhere, surrounding every particle of soil, while roots are much harder to find. Hundreds of metres of hyphae can grow from a single root tip and it is quickly apparent that the roots' function is to nourish the fungi so they can locate food sources for the tree. In return, the fungus gains carbohydrates formed by the pine during photosynthesis. This is why this symbiosis is so successful in the plant kingdom. When the sun shines, the plant benefits from carbohydrate formation but struggles to absorb sufficient quantities of minerals like nitrogen and phosphorus from the ground. By contrast, soil-dwelling fungi have to compete fiercely for scarce carbohydrates. By tapping photosynthesising plants they gain a huge advantage when competing with other organisms in the soil.

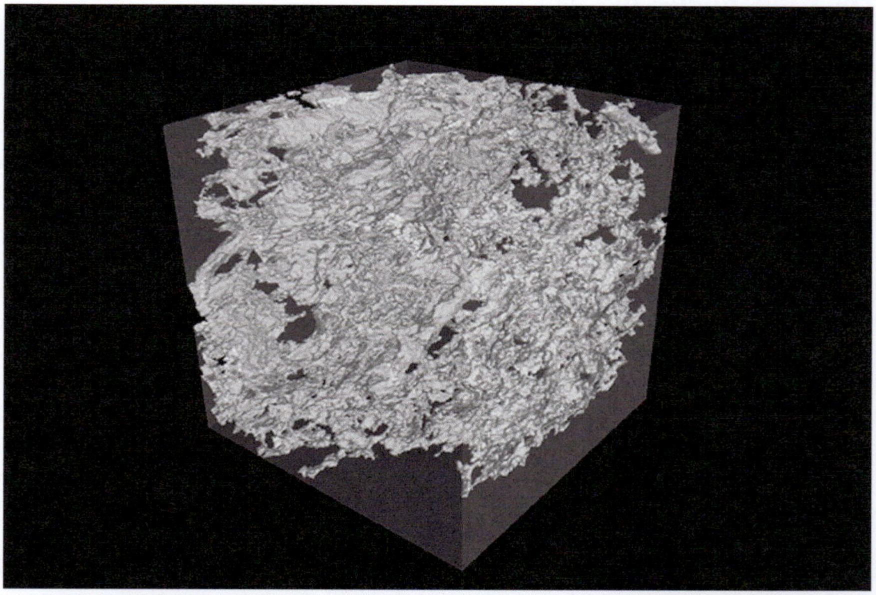

◀ *Underground transport routes.* Using computer tomography, it is possible to generate three-dimensional X-ray images of undisturbed soil (*upper image*) that reveal what soil aggregates and cavities look like. The image shows a cubic centimetre of soil from a field, with all the white areas representing cavities. More than half of the soil volume can consist of pores, but this proportion decreases when soil is compacted by heavy machinery or careless feet in a cultivation bed. Plants find it harder to grow in soil that has been compacted as the air penetrates less easily to oxygen-requiring organisms

Soil that dries out forms small lumps or variously sized aggregates. Organic matter in the lumps absorbs moisture, keeping the soil humid. The cavities between the lumps are full of air, and rainwater moves through these spaces on its way down through the soil. The soil is in constant flux, with transport channels for water and air coming and going as the weather changes

Light, air, water and nutrition. A delicate mycorrhizal fungal mycelium grows from the roots of a pine seedling. The plant is cultivated on a Plexiglas plate with a thin layer of soil to reveal how the mycelium forms the tree's contact interface with the soil. The fungus absorbs the water and nutrition the tree needs and transfers it to the roots. In exchange, the fungus gains access to carbohydrates that the pine stores in its needles with the help of sunlight. All the white areas on the image are fungal mycelium, while individual roots can be distinguished in the centre of the image. In forest soil, a single teaspoon of humus can contain several hundred metres of fungal hyphae. A tree's mycorrhizal mycelium is easy to glimpse if you lift up the moss in the forest, but you need a microscope to identify the type of fungal mycelia that grow from the roots of herbs and grasses

Boundary-spanning collaboration. "Nothing in evolution makes sense except in the light of symbiosis," says symbiosis researcher Lynn Margulis, paraphrasing the classic words of the geneticist Theodosius Dobzhansky, who said, "Nothing in biology makes sense except in the light of evolution." Symbiosis—cohabitation between organisms of different kinds—is common in nature and was a prerequisite for the explosion of biodiversity on Earth. When algae and fungi met and formed lichens, we gained a new group of organisms that, in time, spread to almost all corners of the planet. Growing on the carved handshake in the photograph are crustose lichens

◀ *Nutrient-poor sand.* Grains of sand on the beach easily blow inland on the wind. Farmers in the southern Swedish provinces of Skåne and Halland suffered major problems in the mid-sixteenth century when nutrient-poor coastal sand caked their fields. One of the reasons for this erosion was the practice of using seaweed as fertiliser, which depleted growing conditions for plants such as European beachgrass and lyme grass (*pictured*), weakening their ability to hold the sand in place. Human population growth further increased the pressure on beachside vegetation and the problems persisted until the nineteenth century, when large-scale planting of pine trees took place along the coast

Desertification in Africa

One might imagine that the ability of fungi to bind soil particles together could come in handy in areas susceptible to erosion—so long as one uses plants that attract the right sort of fungi. The Swedish International Development Cooperation Agency (Sida) liked the idea and sent me on a reconnaissance trip to Tunisia, where desertification has increased due to the felling of acacia trees. So I packed my bags with test tubes and soil collection equipment and headed for the airport.

The first African I met was a young Tunisian fellow passenger on the flight to Tunis. A student at Chalmers University of Technology in Gothenburg, he told me Sweden was clean and well ordered compared to his country, where rubbish lines the streets. But he was proud of his roots and invited me to meet his family in Tunis, promising that his grandmother's couscous was altogether superior to the precooked variety one buys in Sweden. Unfortunately, I had to decline this generous offer as I needed to travel south into the desert, but I told him about my fungi and how I hoped they could help to combat erosion. As we bade each other a warm goodbye at Tunis airport I tried to picture a young Swede inviting a stranger from North Africa home to taste his mother's pickled herring or his father's aquavit.

After clearing customs I was met by Hafedh Nasr, North Africa's leading expert on mycorrhiza, with whom I was to investigate how symbiotic fungi might combat soil erosion in the savannah forests on the edge of the Sahara. The acacias in this region are highly resistant to drought, their roots capable of reaching water forty metres beneath the surface. The trees once served as a natural buffer against desert encroachment, but most have been cut down to provide food for livestock and firewood for a rapidly increasing local population. Many were destroyed during fighting between Allied and German and Italian troops during World War II and the forests today are little more than a scattered vestige of a bygone era.

Windblown sand due to inadequate vegetation is a problem not only in the Sahara. The southern Swedish province of Skåne suffered similarly in the early twentieth century, when sand blown inland from beaches rendered arable land infertile. The pine plantations along the coast today testify to how the authorities solved the problem then.

History is littered with examples of how people have been forced to flee infertile land. In *The Grapes of Wrath* and *East of Eden*, John Steinbeck relates how farmers in America's Midwest abandoned their homesteads and relocated to California when the soil on their land was blown away during a period of severe drought in the 1930s. The Midwestern prairie is especially susceptible to wind erosion and giant dust bowls darkened the skies as far away as New York. The soil in the Midwest originated as silt in dry climates south of the Ice Age glacial zone. Known as loess, the silt is finer than sand and easily blown on the wind. Once deposited, it formed thick layers that over time developed into fertile soil. However, ploughing weakens the loess's aggregates, making them prone to collapse under pressure from wind and rain. Thus, loess is suitable primarily for growing grass since this does not expose the ground directly to the elements.

A number of great civilisations, including Genghis Khan's Mongol Empire, have been founded on livestock farming on steppe-covered loess. This soil is common in northern China, where it can form a layer several hundred metres thick, but erodes quickly if used for cultivation—a problem that has escalated in China in recent years due to population growth and rising economic prosperity.

Soil aggregates become increasingly stable in areas of unbroken grassland, where they can lie undisturbed in the ground. Anaerobic conditions eventually take hold in the inner part of an aggregate as mineral particles become packed on the outside and prevent oxygen from entering. Organic decomposition ceases when the microorganisms inside the aggregate are deprived of oxygen, and this process stabilises the aggregate. By contrast, regular tilling or ploughing halts this chain of events and makes the aggregates vulnerable to attrition from rain. If drought follows rainfall, soil particles may be swept up by the wind and blown away. Tilling also damages fungal hyphae in the soil, further weakening the bonds between aggregates.

The lesser of two evils? Iceland has major problems with erosion. The combination of a harsh climate and intensive sheep grazing has left the protective vegetation sparse. Recent years have seen the introduction of American lupin, a plant that makes the soil more fertile by living in symbiosis with bacteria that capture nitrogen from the air. More nitrogen means more plant biomass, which breaks down into organic matter when the plants die. The lupins prevent erosion but outcompete the natural vegetation. Thus the advantages of decreased erosion must be weighed against the drawbacks of reduced biodiversity

All forms of cultivation are negative from a soil structure perspective, but some are more benign than others. The best approach is to refrain from ploughing and instead cover the soil with organic matter. One can also keep the land permanently vegetated by growing crops between rows of trees or by sowing new crops in patches among others approaching their harvest date. All the same, it is sometimes necessary to till the earth so seeds can germinate and grow. Also, unploughed land is a haven for weeds, which often leads to herbicide use, and disease-bearing fungi that thrive on crop residues.

In arid climates, for instance near the Sahara, it is especially important to preserve whatever natural vegetation exists there. Plants tend to be scattered in these conditions and the soil between them is vulnerable to wind erosion. When Hafedh and I journeyed south to Bou Hedma National Park, we saw how the remains of the former savannah woodland provided a sanctuary for wild ostriches and gazelles. Reforestation measures have run into difficulties here because tree seeds often succumb to attack from parasites, while saplings often die of dehydration if drought prevents their root systems from developing sufficiently in their first year. En route to the park, we had to stop to get a permit from the local authorities. After telling the smartly dressed officials about our project we found ourselves shaking numerous pairs of hands before sinking into a comfortable armchair outside the senior official's padded door. Arabic coffee was served and pleasantries exchanged in Arabic and French. I managed to understand a handful of words like *suédoise* and *champignon* but as the hours passed my restlessness increased, along with my frustration at not being able to start work on the project. Eventually it was time for lunch, for which we all decamped to a nearby market. Here my companions bartered hard in finest Arabic style before purchasing the freshest and most tender lamb for delivery to a nearby restaurant—all in honour of their Swedish guest. The chef, apparently unperturbed that his own meat had been deemed not good enough, rekindled the grill and got to work. Plates of harissa appeared and we took it in turns to dip our bread in the hot chilli sauce, easing the spiciness by adding olive oil, according to taste. The lamb arrived on a large platter and we tucked in heartily as the bones, olive stones and breadcrumbs piled up around us. The meal ended with a glass of green mint tea garnished with pine nuts. When it was time to pay the bill I began rummaging for my wallet only to meet a sharp stare from Hafedh. No self-respecting Arab host would ever consider allowing a guest to pay his own way.

Our journey continued through southern Tunisia, the arid landscape punctuated now and then by oases and their dark-green palms and lush vegetable beds. On my mention of the sweet, sticky dates we eat in Sweden at Christmas, Hafedh began talking to the driver and put on his seat belt. We accelerated suddenly, the driver leaning on his horn to tell other cars to give way. A few minutes later we screeched to a halt beside a small truck laden with fresh dates. Our driver hooted and Hafedh motioned to the fruit seller to drive onto the verge. To my ears it sounded like an argument as Hafedh haggled with the seller over the price of a bag of dates while lorries thundered past on the highway. The seller grew increasingly irritated, reluctant to sell the dates, which he apparently did not own. In the end, Hafedh had to settle for breaking off a few dates to give his guest a taste. My curiosity stilled, I

realised I would have to weigh my words more carefully next time. Not for nothing is it said that a good Arab gives his guest food and a roof over his head for three days before asking him what his business is.

We arrived at Bou Hedma National Park in late afternoon to be greeted by yet another welcoming committee. More hands were shaken, cheeks kissed and fresh cups of Arabic coffee served. Feelings of warmth and appreciation mingled with my innate Swedish impatience. We were already behind on our timetable and had spent an entire day in the field without my fingers having so much

◀ *Feeling the urge.* Conducting research in a developing country is very different from doing so in Sweden. Many people inevitably become involved in the project. Some work as drivers or help with sample collection. Others are civil servants from the various public authorities whose permission is required. In certain countries, observers accompany researchers to ensure they abstain from political activities. Everything takes time, but when you meet people with different perspectives and experience new environments, fresh questions are raised and the urge to carry out research gains renewed impetus

The mycelia of mycorrhizal fungi help to bind soil particles into stable aggregates, preventing erosion. The fungi grow better when nitrogen-rich compost is added to the soil. Here we are collecting one of our first samples as part of a project to investigate the extent of fungal presence in the Tunisian desert soil. Since then, the collaboration between Lund University and the Tunisian research institute for agriculture, water and forests has continued to develop and grow

as touched Tunisian soil. Eventually the formalities ended and we were ready to visit Hafedh's research area. Two turban-clad Bedouin carrying spades and pickaxes clambered onto the back of our pickup truck and we sped off on a stony track between the acacia trees. Our first stop was at the edge of the sparse forest, where we wanted to find out if mycorrhizal fungi introduced earlier during the project were colonising the local tree roots—and how far their hyphae extended through the soil. I jumped out of the pickup, eager to start digging, but Hafedh shouted out something about scorpions, whereupon our two companions began hacking at the ground with their pickaxes. Once again, I was forced to stand idly by, until the two men had excavated a hole big enough for Hafedh and me to examine the exposed root structures. We were also accompanied by a group of suited officials of varying ranks, who watched as we decided where to take our first sample of Tunisian soil. I solemnly filled a plastic bag with the soil in readiness for high-tech scientific analysis in Sweden. The hope was that the results would indicate the extent to which fungal presence was helping to improve the soil structure.

Lund University and the National Research Institute for Rural Engineering, Water and Forestry in Tunisia have been working together for many years and our knowledge of the role mycorrhizal fungi can play in improving eroded soil is constantly evolving. Our initial studies of the acacias in Bou Hedma indicated that the fungi which Hafedh had previously introduced to the soil there had been largely ineffectual, both in terms of improving tree survival and soil structure. The fungus, imported from Spain, was unsuited to Tunisian conditions. This led us instead to try to improve the conditions for indigenous fungi already present in the soil, which we did by adding nitrogen-rich compost. This promotes aggregate growth and allows water to penetrate deeper into the earth. Addition of organic matter also aids water retention.

One conclusion from my first African project was that nature is often best at selecting the appropriate species for specific environments. If we set our sights on finding a fungus that can be a silver bullet in combating erosion in different types of natural environment, we are likely to look in vain. A preferable approach is to create optimal conditions for local fungi in their home environments. Nor do I believe the answer lies in genetically modifying organisms to make them more effective. Rather, the way forward is simply to manage land more sustainably.

How Salt Affects Soil

Increasing salinity is another major problem in world food production. It is caused when mineral salts released from the bedrock rise with water to the earth's surface in areas of low rainfall. Salt crystals form at the surface, especially if crops are irrigated with saline groundwater, which can lead to the land becoming infertile. Cato the Elder, who is said to have concluded his speeches to the Roman Senate with the words "I declare that Carthage must be destroyed", sought to achieve this objective by spreading salt on the fields to make them unfit for cultivation. The Romans did indeed take Carthage in the end, but sought to harness rather than destroy the town's resources. They invested in infrastructure, building Roman baths, and when I visited the site, north of Tunis, I found the land to be lush and green. Cato either refrained from implementing his project or else rain washed the salt away.

Annual rainfall is considerably lower in southern Tunisia than in Carthage, and large areas of land here are impacted by salinity. Salt has a number of negative effects on the soil and the plants that grow on it. Sodium, found in common household salt, damages soil aggregates and earth containing high levels of sodium and low levels of calcium tends to become compacted and almost impervious to plant life. Adding calcium, preferably in the form of gypsum, improves the soil structure and adds porosity.

The ways of salt. Some plants manage to grow in saline soils by forming substances that can retain water in their roots even when the groundwater is salty. Some of them secrete excess salt through special glands in their leaves, which has the added benefit of protecting them from being grazed by sheep and goats. The photo shows the salt crystallized into a white powder on the soil surface

No plants can tolerate high levels of sodium ions in their cell cytoplasms. Special ion pumps remove the sodium from the cells, and some plants transport them to other parts where they can do no damage. The grassleaf orache, which grows on beaches in southernmost Sweden, secretes salt from special glands in its leaves. I sometimes chew them on my walks near Laröd, the seaside village near Helsingborg where I grew up. They taste nice enough in small doses but you realise they are unsuitable fodder for animals like sheep and goats, the commonest ruminants in Tunisia.

It is not easy for a plant to find enough potassium ions in soil rich in sodium ions, which is why plants growing in saline conditions often have a potassium deficiency. High salinity also makes water diffuse from plant roots via osmosis—the same process that occurs in human kidneys when we reabsorb water from urine prior to excretion. Thus, plants dry out in saline soil even if they have an adequate water supply.

Salt-tolerant plants sometimes counteract this by forming substances with a high osmotic propensity to help them retain water in their cells. This energy-intensive response enables them to survive in saline conditions. A great deal of funding is going into conventional plant breeding and genetic modification research in a bid to add these properties to modern crops.

In southern Tunisia my colleague Hafedh is carrying out extensive test farming of both domestic plants and imported varieties from salinity-affected areas of North America and Australia, in an effort to create favourable conditions for symbiotic fungi. The main crop we are experimenting on is salt-tolerant barley used for animal fodder. Knowledge about new crops is constantly growing and there is an increasing willingness to experiment. Declining fish stocks in the Mediterranean have caused some fishermen to abandon their traditional livelihoods in favour of farming—with promising results. Successful introduction of foreign plants for high-salinity cultivation contrasts with the failure to combat erosion through the introduction of fungi. But we should not be surprised at the patchiness of our results: many organisms compete to use the soil and we lack knowledge about how to manage the land to ensure that introduced organisms will thrive. We are also largely in the dark when it comes to knowing how species differ in their ability to stabilise soil aggregates. By contrast, we have well-established growing techniques for specific crops and know how best to cultivate the soil and fight the weeds to promote crop growth. Adding compost and other organic matter can be effective in saline soil because a higher humus content helps plants to tolerate salt. We have found in our experiments that mycorrhizal fungi grow best, and hence are better able to absorb sufficient nutrients, when we add compost.

Farming in high salinity environments is a constant balancing act. A type of grass from Pakistan grows well in saline soil but needs intensive irrigation. If you take too much of this water from wells you create a risk of sea water entering the water table and ruining the entire area for agriculture.

Drought-resistant wells. The baobab trees of Sub-Saharan Africa drop their leaves during the dry season to save water. Legend has it that their unusual appearance depends on the god Thora's dislike of the tree, which he threw out of the gardens of paradise. The baobab landed upside down but continued to grow, albeit with its roots pointing skyward instead of down into the ground. When the rains come, the baobab can store thousands of litres of water in its trunk. It is not surprising that the tree enjoys mythical status in some African cultures. The photo shows a baobab tree watching over the cemetery in the village of Fadiout in Senegal

Landslides in Nepal

In Tunisia, wind is the major cause of soil erosion. In the Himalayan kingdom of Nepal, monsoon rains are the prime culprit, weakening the aggregates that bind the exposed soil on mountainsides. A growing human population that needs more

Dehydrating water. The woman in the photo earns a living selling souvenirs to passing tourists near Lac Rose in Senegal. Evaporation is so high due to the heat that salt accumulates on the bottom of the lake, where local people collect it in boats and send it away for refining. Growing crops is difficult here. The high salinity forces water from the plants' roots to even out the difference in salt concentration between cell liquid and groundwater. This is why plants dry out in salty water

land for agriculture and fodder for livestock puts pressure on the natural vegetation. I joined Nepalese colleagues on a visit to a farming community in the Kathmandu Valley, against a backdrop of gleaming snow-capped peaks on the horizon. Villages in the mid-high alpine zone are those hardest hit by monsoon landslides. On our approach to Kathmandu via the main highway from India, the traffic came to a standstill: construction workers were feverishly repairing parts of the road swept away in a landslide. As I scanned the sea of mud from the window of our Mercedes, wondering how we would be able to continue, the driver suddenly hit the accelerator and we found ourselves slithering out of control, skidding left and right, before miraculously emerging from the landslide onto a patch of tarmac. Only after we had resumed our journey did we spot the burnt-out car and bus wrecks down in the valley.

The mountainsides are terraced to prevent erosion but the steepest slopes remain susceptible to landslides. Torrential rain washes fertile soil down the valleys and into the rivers.

Villagers grow corn on the best soil nearest their communities, taking animal fodder and firewood from farther afield. The fields near the villages are fertilised by livestock, while the outlying land is gradually depleted. This results in patchy

Protected soil. Nepal suffers from a major erosion problem related to powerful monsoon rains that wash away soil in areas where forest has been logged and steep mountainsides converted to agriculture. These terraces near the village of Kagbeni in Mustang province are sheltered from the rains, reducing the threat of erosion. The crops are watered by ingenious irrigation channels from a river higher up in the mountains. The village, once an important stop for salt caravans travelling from Tibet to India, is today a popular destination for trekkers. Buddhist monks still use the mountainside caves for meditation

plant growth, exposing the soil to the risk of erosion. Artificial fertiliser would be a solution but developing-world farmers often cannot afford such luxuries. In fact, manure from livestock is not always left in place but instead sometimes used for fuel when supplies run short.

A family in one of the villages invited my colleagues and me to stay in their home of sun-dried mud with a straw roof. In one half of the house stood tethered animals and in the other half, the lady of the house cooked on an open fire. Smoke from the fire impregnated the straw roof and helped to keep pests at bay. We were told that a Swiss aid organisation had donated cooking ovens fitted with chimneys but the villagers had been forced to go back to open hearths because the lack of smoke turned the roofs into a breeding ground for insects. One of the owner's daughters sat on the dirt floor picking corn from dry cobs before grinding them into flour.

Above the village grew a sparse pine forest planted in the 1970s as part of an Australian aid project. The wood, used for timber and pulp, has provided income for the state but not the local community. Villagers initially used pine needles as bedding for livestock, but this led to erosion in areas where the protective layer of needles was harvested. Only remnants of the indigenous forest remain, but they nevertheless contain a variety of useful tree species, some providing firewood and others fruit and a source of traditional medicines. Formerly, the native forest was denser than the pine plantations and the understorey helped to stabilise the soil and protect against erosion. Most of the tree species are suitable for pollarding. When used for this purpose rather than for timber, they provide permanent stability to the soil. Some trees can be successfully pollarded for a hundred years, and expanding this practice in Nepal would substantially reduce erosion and landslides.

Development Aid Meets with Mixed Success

Development aid projects often fail because donors work exclusively through contacts at government level rather than focusing directly on local communities. Yet some reforestation projects have been successful in the Himalayas in recent years. Nepal's forests are state-owned but responsibility for their management is sometimes entrusted to local villages, resulting in better husbandry and the planting of more appropriate tree species.

Through Ecoslopes, a former research project financed by the European Union, I came into contact with a successful village forestry project run by a team of Austrian scientists and local communities in the Kathmandu Valley. Ecoslopes brought together researchers from all over the world to develop ways to use plants to repair and restore areas prone to landslides. Previous initiatives in Nepal had been unsuccessful due to difficulties in getting active cooperation from villagers during spring and summer, when they were busy with farming. The Ecoslopes team addressed this by experimenting with winter tree planting, trialling both native and non-native species with bamboo frames to hold the soil in place and support the saplings. Bamboo is a cheap and widespread resource in Nepal and bunches of canes were fixed in the ground with the help of poles. Many of the saplings survived and also withstood the monsoon.

The trees that fared best were Indian willow (*Salix tetrasperma*) and Nepalese alder (*Alnus nepaliensis*). The latter proved especially suitable for planting on soil recently exposed to the elements, including land damaged by a landslide. Such soils lack nitrogen, a key element needed to form the proteins on which all organisms rely for nourishment. Nitrogen is absent from rock but comprises eighty percent of the air we breathe. The problem is that it takes a lot of energy to turn nitrogen from the air into a form that plants can absorb. A few families of bacteria can perform this task—and it is with these that the Nepalese alder has a reciprocal relationship. The bacteria grow in nodules on the alder's roots and use energy-rich carbohydrates provided by the tree to fix nitrogen. The nitrogen makes the soil more fertile and also leads to an increase in humus when nitrogen-fixing plants colonise the soil. The addition of nitrogen is often critical for successful restoration of the soil because nitrogen disappears when the topsoil is washed away by rain or landslides. Adding organic matter to the soil increases sapling growth and survival rates by benefitting the mycorrhizal fungi that help the trees to absorb

Vital infections. The image shows the nodules that form when the roots of an alder tree are colonised by bacteria capable of capturing nitrogen from the air. The tree provides the bacteria with energy and gains precious nitrogen in return. Alder is an opportunistic species, often first to appear in the wake of avalanches and on soil that emerges when glaciers retreat, and good at stabilising erosion-prone soil

other nutrients. One spin-off from Ecoslopes is a database detailing the properties of different tree species and how they can be used to stabilise erosion-hit areas internationally. Certainly, one must hope that initiatives of this sort will become more widespread in Nepal in future.

Is Organic Farming Sustainable Agriculture?

Sustainable agriculture means a system in which nutrients circulate and are returned to the soil in equal measure to the nutrients removed during harvesting. Nitrogen-fixing plants can play a key role in this context by obviating the need to fertilise the soil with nitrogen. The legendary Swedish author August Strindberg knew this when he wrote *The People of Hemsö* in 1887. In the book, Madame Flod invites Carlsson to visit her farm on the island of Hemsö to educate her son, Gusten.

> Gusten explained that harvesting land year in and year out depleted the vitality from the soil, which needed to rest, just like Man. But Carlsson gave a correct, if somewhat hazy, explanation of how the clover crop fertilised the earth rather than depleting it, and kept it free from weeds.
>
> 'Well, I'd never heard of crops fertilising the soil,' replied Gusten, who could not understand Carlsson's scholarly explanation of how grasses took most of their nutrition 'from thin air'.

Pesticides and artificial fertilisers are not used in organic farming. The focus is on confining the nutrient cycle to the farm, for instance through keeping manure-producing livestock. Animals are fed a mixture of grass and herbs that not only

More with less. Biodiversity increases if we farm organically and avoid pesticides and artificial fertilisers. Poppies and cornflowers can survive on field margins, while grass and other species that benefit from nitrogen-rich artificial fertiliser become less dominant

provides excellent fodder but also fertilises the soil with nitrogen-fixing plants like clover. The roots of clover and other plants can fix several hundred kilograms of nitrogen per hectare. The findings of a major Swiss study comparing the environmental consequences of conventional agriculture with organic farming are unequivocal: over a period of twenty years, organic farming led to more fertile soil with a superior structure, higher humus content and greater diversity of plants and terrestrial animals than conventional farming. On the other hand, crop volumes were on average one fifth lower on organic farms. Potatoes fared especially badly in the study, with blight causing major losses. Potato blight caused the Great Famine in Ireland in the mid-nineteenth century and potato growers have yet to find a silver bullet. Some strains of potato developed to be resistant to blight have later lost their immunity.

Organic farming involves the sustainable use of resources. But the fact remains that were it not for artificial fertiliser, the global population would be much lower than it is today. There is a broad consensus that the number of people on the planet will stabilise at around nine billion by 2050. If we are to feed all these mouths, we will need to use all the means at our disposal to increase world food production.

But over-intensive agriculture causes serious problems. Western countries import animal feed made from soybeans grown in vast monocultures in countries like Brazil. Since monocultures are vulnerable to disease, the crops must be intensively sprayed with pesticides, using chemicals long banned in Sweden. The pesticides make farmers ill and the land also suffers from such an intensive, monotonous form of agriculture. Yet there is hope in Brazil, where organic farming has become established in the southern part of the country and is gradually gaining ground. Trials in agroforestry—mixing silviculture with annual crops—have proved successful. In agroforestry, the ground is never left bare and because a wide variety of different trees and crops can be grown, diseases take longer to spread, which means farmers can use fewer pesticides than in monocultures. Growing trees alongside other crops is a method that has come farthest in permaculture, which is not only an agricultural technique but an entire philosophy. Permaculture developed in Australia in the 1970s and has since spread worldwide.

When I was young I worked as a volunteer at Skillebyholm Farm near Järna in central Sweden. The farm held courses in biodynamic agriculture and I met many enthusiastic farmers from all across Europe. Biodynamic agriculture differs from organic farming as it treats the farm as a closed loop that is replenished with manure from farm animals, which means livestock are always part of a biodynamic farm. Like organic farming, biodynamic agriculture avoids artificial fertiliser, pesticides and herbicides. It also features some anthroposophical ideas, regarding cosmic radiation as a special life source and suggesting that cosmic energy can be harnessed in biodynamic products made in special conditions. I have myself helped to dig up cow horns filled with fertiliser under the light of a full moon to prepare substances to stimulate the composting process. Anthroposophical theory also holds that the soil should sometimes be left bare to enable cosmic radiation to have full effect. But in my opinion, vegetation is a skin that shields the soil from wind and rain. If we have to till the soil, we should sow seeds as soon as possible or cover it with organic matter. In the next chapter we will discover that it takes a long time for new soil to form under natural conditions, but that we can speed the process up by adding organic matter to areas suffering from landslides or desertification. We would be able to stop erosion if we added sufficient organic matter to raise soil humus content.

Further Readings

Brady NC, Weil WR (2008) The nature and properties of soils. Prentice Hall, New Jersey

Hyams E (1952) Soil and civilization. Thames and Hudson, London

Lundberg J, Moberg F (2009) Ekologiskt i Brasilien. Deltagardriven certifiering och lokal handel för ett hållbart lantbruk. Naturskyddsföreningen

Mann CC (2008) Our good earth. The future rest of the soil beneath our feet. National Geographic, Sept 2008

Mollison B (1991) Introduction to permaculture. Tagari Publications, Australia

Mäder P et al (2002) Soil fertility and biodiversity in organic farming. Science 296:1694–1697

Ritz K, Young I (2004) Interactions between soil structure and fungi. The Mycologist 18:52–59

How Soil Is Formed

The flakes of inky thunderclouds
Above the convex whiteness of the glacier.
A lifeless river dressed in yellow till
And sun rays, glistering like spears at daybreak,
Pierce through the silent, silver sky
"Awake, oh Earth, awake!"—I cry
"Break free from this eternal coldness!

Transform your till and gravel into soil
And let the frigid puddles of the tundra
Be overgrown by reeds, green and unspoiled!"
The Earth is silent, and the pearly slabs of glaciers
Lie far beneath the icy sun, as in a trance.
Indeed, no human measure ever
Can wrap around this ancient, vast expanse.

N. Zagorskaya

Abstract Plants play a crucial role in the formation of soil; soil formation accelerated when the first plants wandered onland 400 million years ago. One of the conditions enabling plants to develop was that algae joined forces with fungi, which were better at absorbing phosphorus from the ground. However the first soils were formed by microorganisms. Their complex miniature worlds, so-called biofilms, are described and compared with human cities. Acids secreted by microbes and plant roots accelerate the weathering of minerals, and dead plant material helps to build up humus in the soil. But soil formation is a slow process, and one has to be ingenious to follow the phenomenon over long periods of time. You can take soil samples where glaciers are receding, and along the gradients from the coast towards the highest shoreline, in regions which are in a process of post-glacial rebound still rising after being depressed under the weight of the ice during the last ice age. This makes it possible to follow the soil formation process over thousands of years.

H. Wallander, *Soil*, DOI 10.1007/978-3-319-08458-9_2

The Very First Soils

How the first soil formed on our planet remains a matter of speculation, but we do have a few clues. Paracelsus, a sixteenth-century Swiss scientist and alchemist, made interesting observations of paper-like fragments that he found while hiking in the Alps. When he put them in water they swelled into a slimy, jelly-like mass containing a complex microbiological world later compared to the structure of human cities. In water, this gelatinous material helps prevent the organisms from being washed away by currents. One could liken it to taking shelter indoors from a storm. The material also provides insulation, safeguarding the organisms from excessive temperature swings. Waste products are transported in channels akin to urban sewers. Remarkably, the organisms in these societies can communicate with each other. Each bacterium emits signalling molecules whose concentration determines when the bacteria start producing jelly.

Blue-green bacteria play a critical role in these gelatinous worlds—or biofilms as we call them today. Paracelsus named them *Nostoc* because their mucus reminded him of the symptoms he experienced when he caught a cold. Nostoc has the same ability as the bacteria in bean and alder roots to bind nitrogen in the air to proteins.

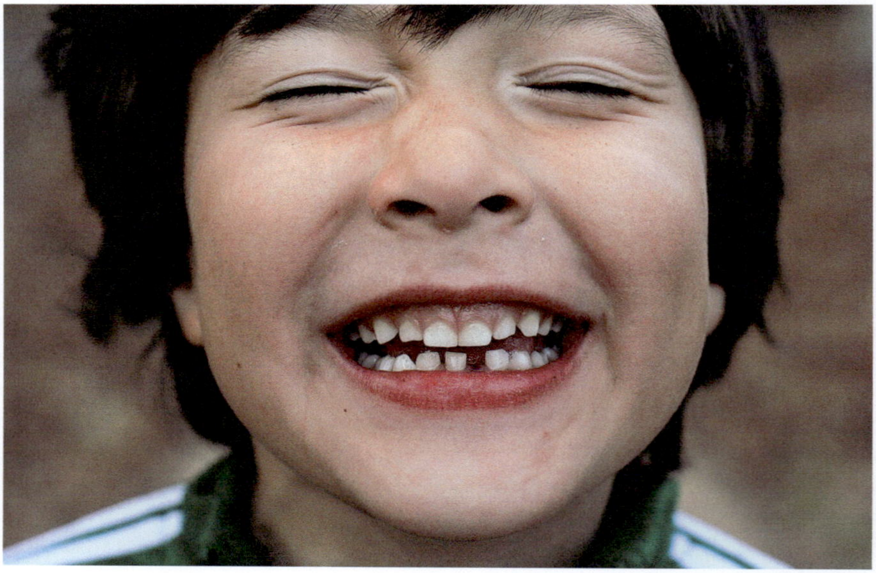

The first soils. Even if we brush our teeth morning and night, bacteria are continually forming new colonies inside our mouths. The oral cavity is a favourable environment, with regular provision of nutrition and just the right temperature, and teeth are a suitable material to adhere to. Once the bacteria have attached, they form a protective material in which they embed themselves—known as biofilm. A similar process occurs on the roots of plants and on soil particles where there is good access to nutrients. The very first soils, predating plants and lichens, were formed by biofilm material gradually gathering in the ground and becoming organic matter

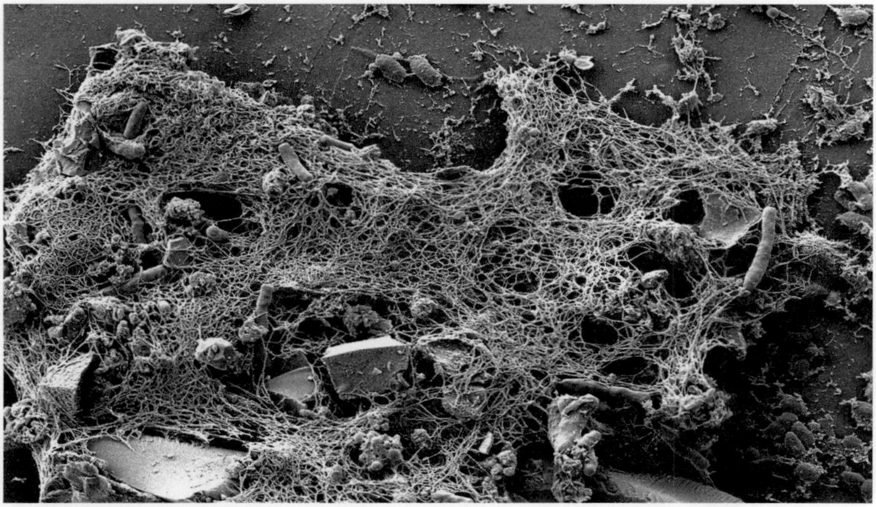

Supermuck. Bacteria in a biofilm form a gelatinous material that protects the colony from external disruption. In human lungs, for example, antibiotics can have a hard time reaching the bacteria in a biofilm, which can make some infections hard to combat. Observing the true nature of a biofilm is difficult. In an electron microscope, the gelatinous material dries into a spaghetti-like network. The image shows rod-shaped bacteria embedded in the material

Like these plants, Nostoc can photosynthesise and form carbohydrates from sunlight and carbon dioxide. In South Africa and Australia, scientists have found biofilm traces more than three million years old—the oldest known life forms on the planet.

Biofilms often form on underwater rocks. We have all trodden on submerged rocks at the seaside and can doubtless imagine how fragments of biofilm might dry out at low tide and be blown by the wind onto the shore. But some biofilm organisms were able to survive on land. Many fungi have special pigmentation to protect them against ultraviolet light, and their ability to withstand drought by secreting a tissue-protecting barrier with high drought tolerance is likely to have played a critical role in the formation of the first soil. Today we know that most areas of land exposed to the atmosphere are covered in biofilms, some invisible to the human eye and others forming delicate patterns on old stone buildings.

Inspiration from Crete

I took the opportunity to study fungi and their unique properties on a walking holiday one summer in Crete. The terrain on the southern side of the island is so rugged that there are no coastal roads: getting there means walking or taking a boat. The trail that runs west from the village of Sougia, south of Chania, passes a harbour and then enters a leafy ravine where pink oleander bushes bloom and olive trees offer welcome shade. Pigeons and tree sparrows nest in the steep sides of the gully—an environment

not dissimilar to the urban environments that these species frequent today. After about half a mile you can turn left into a pine forest which leads to a ridge with a view of the sea. The scent of thyme floats over the thorny scrubland and wheatears flash their white rumps as they fly from stone to stone. A little farther on you reach a spectacular view

◀ *Splendid life artists.* Growing on rock is an arduous business. Temperature varies greatly between day and night, periods of drought may be prolonged and the sun's ultraviolet rays can damage cells. But if an organism can adapt to such an environment, it is often left in peace and not outcompeted by other species. Some fungi form water-repellent substances that minimise water loss. They also secrete acids that corrode rock so they can penetrate it with their hyphae, gaining protection. Black yeast fungi have lived in this way since prehistoric times. Four hundred million years ago, some of them combined with algae to form lichens—one of nature's most successful organisms. Lichens have spread to most environments on Earth and display an array of colour and form. Here the elegant sunburst lichen forms distinct and well-delimited patches on a gravestone near a medieval church in Vågå, Norway

over a bay immersed in green in the valley below. This is Lissos, once a Roman commercial centre and now an archaeological attraction with an ancient temple and ruins of Roman houses and storerooms. The finest relics are on view in Chania Museum but the ground is still peppered with marble fragments, perhaps the remains of an ancient column. It is hard to believe the area has not been cordoned off to the public but it makes the excursion all the more exciting—a bit like being on a treasure hunt.

The marble shards I found were peppered with tiny black dots made by a yeast-like and highly durable fungus. Even if washed from the surface, the fungus soon grows out again from the inside. Black yeasts are among nature's great survivors and can adapt to severe drought and temperature fluctuations. They are found from the most inhospitable deserts to the coldest places on Earth, their black pigment protecting them from ultraviolet radiation. They are extremely adept at penetrating between the mineral granules inside a stone, where they create cavities that protect them from drought and the rays of the sun. Their only source of nutrition is wind-borne particles that adhere to the surface of the stone. Black yeasts are regarded as predecessors of lichens, organisms that have successfully colonised most environments on the planet. Lichens are a symbiosis between a fungus and an alga. The fungus provides a protective layer that enables the alga to photosynthesise even in the dry conditions found on rocks and other exposed places. Scientists believe that when the Earth was in its infancy, the action of fungi caused rocks to crumble and mingle with dead fungi and algal tissue to form the first soil. Some soils are formed in this way even today—that is, without any intervention from plants. In Antarctic deserts, which are too cold to sustain plant life, soil formation is based on blue-green bacteria, fungi, algae and lichens that penetrate a millimetre or two into rock surfaces, thereby causing the stone to slowly fracture and join dead microorganisms in forming a primitive soil. There are other places where one can study soil formation. Come with me on a journey to some of them.

Soil Formation on a Glacier's Edge

A flight over northwest America en route to Seattle gives a spectacular view of Mount Hood, which stands in the middle of the Cascade Volcanic Arc, a mountain range that saw a major eruption less than thirty years ago when the top of Mount Saint Helens exploded with such force that it deposited ash residues as far away as Costa Rica. We know this due to scientific studies of soil formation in the Costa Rican rainforest, where leaves and other matter that collects in tree forks gradually forms a soil base in which seeds can germinate. It was several centimetres down inside the soil of one such fork that researchers found ash originating from the Mount Saint Helens eruption.

On arrival at Seattle airport, I am greeted by American colleagues who are to take me to Lyman Lake, which lies at an altitude of eighteen hundred metres near Glacier Peak Wilderness National Park—a journey of many hours by car and then boat. This pristine area was formed by meltwater from a glacier that has retreated at a rate of eight to twelve metres per year during the last century and now lies some distance from the water's edge. We are here to study pristine soil never previously exposed to the elements and answer a host of questions, such as the impact of rainwater when it seeps through the soil, what species of plants colonise the area and what happens to the soil when vegetation takes hold. Thanks to photographs, we know the exact position of the glacier's edge and how it has shifted over the last hundred years.

On our way to the lake, we hike through dense deciduous forest that gradually turns into sparse fir woodland. Yellow wolf lichen grows on the boughs of the

Tentative steps towards new soils. When a glacier melts and recedes, it leaves behind an assortment of fine and coarse soil particles—everything from clay to rocks. Known as moraine, this mixture slowly becomes soil due to the action of the elements and as plants take root. The stony moraine contains plenty of phosphorus and potassium but lacks nitrogen. Plants must therefore find another way to access this critical element. The yellow plant in the foreground—alpine birdsfoot-trefoil—lives in symbiosis with bacteria that capture nitrogen from the air. The nitrogen content in the ground accumulates when these plants decay, facilitating the establishment of other plant species

trees, whose elongated crowns necks remind me of Donald Duck's national park visit with his nephews. As we approach the glacier, the forest thins further and the blueberry scrub grows higher.

A cold draught from the glacier hits us as we approach. Countless rivulets of murky meltwater run from the ice, forming a stream that continues down into the lake.

As it moves down the slope, the glacier slowly grinds underlying gravel and stones into smaller pieces. The ice is in constant forward motion but will retreat at the edges during persistent or accelerated periods of melting, as during summer. The edge of the ice will recede if this crimping exceeds the distance the glacier travels during the year. If the ice sheet ceases to move the glacier becomes dead-ice and after it melts, it leaves behind an expanse of pulverised rock that eventually forms the basis for soil. Soil is the top layer of earth that is exposed to the elements and supports life forms. It begins to form as soon as vegetation takes hold. It can be partially unsorted, as with moraine deposited in front of a glacier, or more finely sorted into fractions that are carried away in meltwater as silt. The silt settles as mud or sand at the bottom of the lake and forms muddy or sandy soil when the lake eventually dries out.

Close to the glacier, the loosely piled rocks are treacherous to walk on. In winter the water freezes, dislodging particles of gravel and sand that in spring are carried by rainwater into the crevices, where they stabilise the more outlying rocks. This is the start of soil formation—a process that begins when clumps of grass and sedge take root thirty metres or so from the edge of the ice. They do not look like much, but when they die they provide the organic matter needed for willow bushes to grow.

Return of the trees. The glacier in the photo has retreated by an average of ten metres per year over the past century. Various plants have gradually gained a foothold: lichens, mosses, sedges, small willow shrubs. After eighty years, soil formation has reached a point enabling the fir trees in the foreground to take root. It is fascinating to think that the entire Nordic region has undergone this process

It takes around eighty years and the arrival of the first blueberry and other low shrubs and conifers for soil formation to gather genuine pace. The low air temperature slows the decomposition process. Blueberry leaves and conifer needles also contain components that degrade slowly, resulting in an accumulation of organic matter at the surface. Dissolved organic compounds seep into the soil with rainwater and dissolve the ions from mineral particles. The vegetation absorbs some ions, while others collect further down in the soil or leach into the groundwater. As time passes, the soil separates into an upper leached layer and a lower enriched layer, though this process can be affected by the presence of earthworms, whose burrowing mixes the soil profile. At Lyman Lake, one can study the soil formation process during the hundred years immediately following the deposition of moraine by the glacier. There are plenty of other places to visit for those who wish to study the process over a longer time frame.

Podzolisation and Soil Formation in Sweden

In Sweden, the ice sheets melted around ten thousand years ago in a process mirroring that at Lyman Lake. In their wake, they left an unconsolidated moraine that is the dominant soil type in Sweden. During the Ice Age, the weight of the ice depressed the surface of the Earth's crust, which is now in a process of post-glacial rebound as the land regains its former form. In northern Sweden, the land is rising at a rate of one centimetre per year and here, as at Lyman Lake, we can use a natural time experiment to study the progression of the soil formation process, this time under Scandinavian conditions.

I travelled to Åkerbäck, some twenty-five kilometres inland from the Baltic coast in northern Sweden, to take some soil samples. A nearby patch of shingle beach reveals the line the coast took around eight thousand years ago. Lush spruce forest has dominated the landscape for the last three thousand years or so, having colonised the area from the north. So stony was the ground that it was impossible to use my soil sampler, though after much effort I managed with my colleague's help to dig a deep enough hole to examine the soil structure. Above the mineral soil, which was primarily moraine, lay a ten-centimetre layer of organic matter topped by newly shed spruce needles and leaves embedded in blueberry twigs and moss. The matter adjacent to the mineral soil had degraded into a black paste-like substrate showing no direct evidence of needles or leaves. The upper section of the mineral soil was greyish-white in colour, becoming reddish lower down.

Carbon dioxide emitted by plant roots and microorganisms dissolves in the soil to form carbonic acid, which accelerates the disintegration of soil minerals. The acidification effect is augmented by the positively charged hydrogen ions that these life forms release in exchange for the potassium, magnesium and other positively charged ions that they need in order to grow. As the most easily weathered minerals disintegrate and leach, the soil looks bleached because only grey quartz remains and is resistant to disintegration. The bleached soil is called the eluviated

A postglacial profile. Podzol is the most common soil type in Sweden, formed when conifers grow on acidic bedrock, such as granite, in places with abundant rainfall. The bedrock and acidic conifer leaf litter result in slow decomposition, leaving organic material to accumulate on the ground. Acids from the organic layer are carried in rainwater and dissolve the easily weathered minerals in the top layer of soil. A bleached soil is thus formed in which only grey-white quartz is left. Iron and aluminium from the bleached eluviated soil layer trickle down and form rust-red oxides in the layer below, which becomes the reddish illuviated horizon. Below this, the original moraine has not yet been affected

horizon and at Åkerbäck it reached a depth of twenty centimetres. Below it was a reddish layer known as the illuviated horizon. The reddish colour comes from iron and aluminium dissolved from the upper soil layers, a process aided by the seepage of dissolved acids from organic material lying at the surface.

Podzol is commonly found in Swedish coniferous forests and forms on acidic bedrock where the decomposition process is slow and in areas of plentiful rainfall that accelerates the leaching of minerals to the lower layer of soil.

◀ *Earthy hues.* Soil gets its reddish tint from iron and aluminium oxides. Known as ochre, these compounds have been used as pigments since ancient times. Aboriginal Australians still use ochre to colour their faces and for their traditional bark paintings. During the 1800s it was popular to extract ochre for textile dyeing. The southern French village of Roussillon was an ochre centre and the warm colours of the buildings there still bear witness to that period. Nowadays, the village is a tourist destination where visitors can wander round the old open quarry and enjoy the sunlight playing on the colours in the soil and vegetation

Coniferous forest litter promotes podzolisation and the phenomenon is most strongly observed in New Zealand in areas with high precipitation, where the kauri tree grows on acidic ground. The tree sheds large fragments of bark, which intensifies podzolisation in the area of soil directly beneath the tree, resulting in egg-cup-shaped deposits of bleached soil in podzolised areas.

Time is a key factor in soil formation. At Åkerbäck we observed how both bleached and rusty soil layers grew thinner the nearer to the coast we did our sampling. A few hundred metres from the shore, it was hard to identify any bleached layer at all. On the shoreline itself we found a cordon of alder trees, whose nitrogen-fixing root bacteria had helped bind nitrogen in the soil—a critical factor enabling spruce to colonise.

Plants Accelerate Soil Formation

How, one might well ask, would the soil formation process work if plants were absent? Rainwater washes away soil on an unvegetated hillside but penetrates flatter ground, possibly allowing eluviated soil layers to form because atmospheric carbon dioxide dissolves in rainwater and makes it acidic. But it would doubtless take time.

In summer I often stop at a place called Kastad, near Gjövik in Norway, on my way to my cabin in the Norwegian fells. There is a clearing in the spruce forest where the ground is completely bare. Clumps of grass grow around the edges amid a string of low, stressed spruces. The soil here is highly polluted by lead—not from human activity but from an underground source of galena, the natural mineral source of lead sulphide. On top of the galena lies a metre-thick layer of moraine. The area is interesting because the barren space allows one to study the significance of plants for soil formation. Eluviated soil layers always occur in soil where spruces have grown—soil where apatite has leached from the topsoil. Apatite is a phosphate mineral and important source of nutrition for the spruce, and the fungi that live in symbiosis with the tree's roots are effective at utilising it. Apatite dissolves from the topsoil due to the action of rainwater on the bare ground, though our models indicate that the rate of weathering is one hundred times faster when spruce grows there. Clearly, ground vegetation plays a key role in soil formation, which helps explain why soil in the Antarctic, where vegetation cannot grow, is so thin and underdeveloped.

A pocket in time. Take a walk through the spruce forest near Gjøvik in Norway and you may come across this clearing, devoid of vegetation due to the high lead content in the ground. The area was discovered by Norwegian geologists who ascertained the presence of mineral deposits by noting changes in the vegetation. Below the surface is a seam of galena which may be mined in the future. Scientists use the area to study how vegetation affects soil formation, comparing soil in the forest to that in the clearing

Nature's Own Soil Archive

If you want to study soil formation over a long period of time, the place to go is the Franz Josef glacier in Westland National Park on New Zealand's South Island. The area receives high annual snowfall of three to five metres in liquid equivalent, which has resulted in one part of the glacier almost reaching the coast and a milder climate zone. The temperate rainforest just below the glacier edge stands in sharp contrast to the wall of ice. Franz Josef has left deposits along the narrow valley that leads to the sea and has retreated many kilometres since the Ice Age. Uniquely, the glacial deposits here date back to the last interglacial period, around one hundred and twenty thousand years ago, allowing scientists to study soil formation over a much longer time frame than in areas where glaciers are shrinking.

Soil composition is ever-changing, with no end point at which one can say the soil is "finished". In fact, it is preferable to view soil formation as a two-step process consisting of an accumulation phase and a depletion phase—a process that begins all over again when a new ice age arrives. This is because the ice scrapes away the old weathered soil, paving the way for soil formation in the moraine left behind when the glacier melts. Nitrogen-fixing plants tend to

Between the ice. It snows copiously on the Franz Josef glacier in New Zealand—three to five metres per year in melted form. The high precipitation affects the moraine left behind by the melting glacier, weathering it faster than on many other glaciers. The rainwater slowly dissolves the phosphorus in the soil and carries it away. Moss grows close to the glacier edge and a temperate rainforest covers the valley. However, soil phosphorus decreases with distance from the ice, eventually falling to such a low concentration that shrubs replace the rainforest

establish a foothold during the accumulation phase, building up the earth's nitrogen content. Phosphorus content is high at this point but then enters a steady decline. This is because the presence of plant roots makes the soil acidic and results in phosphorus leaching from minerals. If not absorbed by plants, the phosphorus may be washed away by rain. It can also attach to aluminium and iron-rich surfaces, which also takes it out of the equation. As phosphorus content declines, plants start to economise on its use, retaining their leaves for longer and adapting to a lower level of nutrients. This adaptation comes at the expense of reduced growing power stemming from reduced efficiency in photosynthesis. At this point, the vegetation will thin and once-tall forest may revert to scrub. The rate of this process is determined by rainfall. At the Franz Josef glacier, it has taken around twelve thousand years for phosphorus leaching to affect plant growth. The result is an increase in coniferous trees and later on a dominance of species-poor but stress-resistant scrubland. Soil formation over even longer timeframes has been studied in some areas. Studies spanning several million years are possible at Hawaii's lava flows, where the transition from nitrogen to phosphorus deficiency can take several hundred thousand years due to lower rainfall than in New Zealand.

The First Plants Needed Fungi

Phosphorus deficiency also played a key role when plants evolved from exclusively aquatic to terrestrial organisms. The first plants grew on the sea line and had creeping rootstalks, or rhizomes, rather than roots. Aquatic plants easily absorb phosphorus dissolved in the water, but phosphorus on land is fixed and hard to access. The plants that successfully absorbed sufficient phosphorus in this inhospitable environment gained a competitive advantage in terms of stronger growth and a better survival rate. Here, they had fungi to thank. The fungi attached themselves to the plant to gain access to carbohydrates. When confronted with a phosphorus deficiency inside the plant they absorbed phosphorus from the soil, some of which became a food source for the plant. Fungi that did not kill their host plant gained continuous access to carbohydrates, giving them an advantage—and by transporting phosphorus to the plant, they ensured a stable supply of carbohydrate for themselves as photosynthesis requires phosphorus. This was how the mycorrhizal symbiosis, one of the natural world's most successful collaborations, evolved. Mycorrhizal fungi exist on between eighty and ninety percent of plants today. Terrestrial plants would probably have not evolved were it not for the presence of collaborative fungi.

Conditions for soil formation changed significantly when terrestrial plants developed around four hundred million years ago. The plants produced vast amounts of biomass that over time created the organic matter needed for soil to develop. As mineral soil weathered, so the soil formation accelerated.

Ancient recipe for success. During the dinosaur era, gymnosperms such as gingko, cone palms and, later, conifers dominated the flora. Cone palms succeeded by joining forces with cyanobacteria capable of capturing nitrogen from the air. Such symbiosis is today found in some lichens but is unusual among plants. One example is the *Gunnera* family (photo), which is among our oldest angiospermous (flowering) plants and which existed at the time of the dinosaurs. Other groups, including legumes and alders, went on to evolve collaborative strategies with different types of nitrogen-capturing bacteria

Today, it takes a long time for new soil to develop in areas where the topsoil has eroded—too long to restore the areas we have destroyed. The best way to speed up this process is to plant nitrogen-fixing plants that bind carbon and nitrogen in the soil as this creates conditions for other plants to establish themselves and for organic matter to accumulate, in turn creating stable soil that is less prone to erosion. In the next chapter we will explore how the soil's properties determine what grows in it—and how the way we manage the land plays a part in determining what plants will grow there.

Further Readings

Brady NC, Weil RR (2008) The nature and properties of soils. Prentice Hall, New Jersey

Gorbushina A (2007) Minireview, life on the rocks. Environ Microbiol 9:1613–1631

Lambers H et al (2008) Plant nutrient acquisition strategies change with soil age. Trends Ecol Evol 23:95–103

Smith S, Read D (2008) Mycorrhizal symbiosis, 3rd Edition. Elsevier Academic Press, San Diego

Floral Splendour and Shortage of Nutrients

At midsummer the flower was herself,
her wingbeats the essence of summer.
From sprout to seed she flew
on red poppy wings.

Harry Martinson

Abstract The level of nutrients in the soil is of great significance to the composition of the vegetation. In this chapter I discuss why calcareous soils have a greater diversity of species than acid soils in Europe, whereas in Australia the opposite is true. One of the explanations to the great species diversity among plants is that there are many ways of adapting to the nutrient levels in the soil. Mycorrhizal fungi in the soil can also be a determining factor for diversity, but the conditions are different in Europe and America. During my walks in Hörjelgården, a farm where land is used in the same way as in the 1700s, I reflect on why the agricultural landscape has become increasingly species-poor. I also consider the adaptations of plants during hikes in the mountains and on the limestone plain of Öland. Sometimes, plants can indicate the presence of metals in the ground and there are several examples of how they have helped geologists to locate valuable ore. The chapter concludes with an example in which a beautiful ridge landscape is exploited and converted into an open sand quarry. When the sand quarry is later abandoned, it turns into an attractive area for solitary bees, where children's games create the conditions for the bees' survival.

Carpets of Flowers in Sarek and on Öland

When hiking in Sarek National Park, majestic mountains with evocative names like Akka, Sarvatjåkkå and Pierikpakte surround you. For all the grandeur, there is scenery on a smaller scale that is also worth watching, though you may have to look carefully. After hiking for hours, the white petals of a mountain avens suddenly appear at your feet. Is the flower growing here by coincidence or because the soil is different?

H. Wallander, *Soil*, DOI 10.1007/978-3-319-08458-9_3

Plant distribution hinges on multiple factors including climate and, at Sarek, grazing by reindeer. Most plants have similar requirements in terms of nutrients, and it is unlikely that the mountain avens has specific needs not shared by other

◀ *Necessity—the mother of invention.* Phosphorus is a vital plant nutrient. In calcareous soils, plants find it hard to obtain phosphorus because of its affinity for calcium, to which it binds closely. Plants have sought to overcome this through various evolutionary adaptations, and calcareous soils are often rich in flora as a result. This diversity can be observed in places like Sarek national park in northern Sweden and on the Baltic island of Öland, which lies on calcareous bedrock

species. But access to nutrients in the soil can be variable and plants adapt to this intermittent supply in different ways. In all probability, there is something in the soil that enables the mountain avens to outcompete other plants in this environment. A common misconception exists that different plants have different nutritional needs. Just visit your local garden centre and look at the array of different fertilisers on offer. Lemon trees, we are told, need one kind of fertiliser, vegetables another and lawns yet another still. Much of this, though, is sheer marketing designed to encourage us to buy more products.

One day in early August I stepped off the bus at the Sourva hydropower dam near Stora Sjöfallet National Park, which lies just to the north of Sarek. Accompanying me were Johan and Joakim, two hikers highly experienced in exploring this expanse of wilderness. Our first job was to divide the camping gear between us. I drew the task of carrying the four-kilogram three-man tent and Johan the ultra-light portable stove. Joakim carried arguably the most valuable cargo in his head: his intimate knowledge of hiking in Sarek. It took us two days to reach Pielavalta, the high plateau in the centre of Sarek National Park and site for our base camp. It was my turn to cook the evening meal and, to my friends' surprise, I produced a fresh egg. Who knows, maybe I was the first of Sarek's many weight-conscious hikers to carry an uncooked egg halfway across the park. But I needed it to cook the lentil burgers I planned to make that evening. I let the dried corn and green lentils soak in the hot water before mixing in the egg and instant mash potato and frying the patties in olive oil. After a long day of hiking we all ate heartily.

The day after we visited a magnesite quarry on the northern face of Ähpar, above Lake Vuoinesluobbalah. Five hundred tonnes of magnesite were mined here during World War II to make fire-resistant stone linings for steel kilns. The quarry was abandoned after six months due to a plane accident—a mishap that saved Sarek from further development.

I became interested in the quarry, located in the most botanically interesting area of Sarek, after reading about it in Swedish writer and photographer Claes Grundsten's books. Magnesite consists of magnesium carbonate, an unusual mineral that resembles calcite (calcium carbonate). Both calcite and magnesite affect soil conditions in similar ways, their carbonate turning water in the soil alkaline by binding acidic hydrogen ions.

To reach the magnesite quarry we left Pielavalta and hiked through Pierikvaras, a low pass beneath the cloud-shrouded Mount Slugga, a peak that looks the same from all angles. In the distance the quarry shone white on the horizon, presumably the result of vegetation being shorn away when mining started here. We found an abandoned stone saw and a few neatly stacked slabs of sawn rock: relics of a bygone age.

At first glance, the vegetation along the streams running down the hillside from the quarry looked no different from what we had encountered in the previous two days. This disappointed Joakim, who was hoping to learn more about Sarek thanks to my botanical expertise. But after taking off my backpack and examining the ground on hands and knees I found a few knobbly branches growing close

◀ *The flytrap*. Plants that live in extremely nitrogen-deprived environments must adapt to survive. The sticky leaves of this alpine butterwort have captured insects that the plant's enzymes will slowly dissolve to provide a much-needed nitrogen supplement. Another carnivorous plant is the marsh-dwelling sundew

to the surface, the leathery leaves and aromatic scent confirming the presence of *Rhododendron lapponicum*—one of only two wild rhododendron species in Sweden. In due course I found Alpine arnica, Arctic bellflower, and Arctic white heather, three calcareous soil species. Joakim's mood began to lift. The highlight came when we found Alpine butterwort, an insect-eating plant that evolved its carnivorous habits as a dietary supplement in nitrogen-poor soil. Alpine butterwort is a low-growing plant with sticky leaves which trap insects and slowly dissolve them with digestive enzymes. It is said that you can make sour milk from alpine butterwort if you mash the leaves with a wooden spoon and add milk.

Eagle eyes are sometimes needed to spot nature's nuggets. But the more one understands the interplay between soil and plants, the more one appreciates the subtle changes that occur—changes that generally elude the layperson.

I learned a lot about horticulture when I spent a summer at Capellagården School in the village of Lilla Vickleby on the Baltic island of Öland. While I pottered about in the vegetable patch, the other students would paint watercolours of the unique Öland landscape, and the shifting colours of their work reminded me of the light that Claes Grundsten so ably captured in his Sarek photographs. The plaintive whistle of the golden plover can be heard in both places, and while the Montagu's harriers that frequent Öland's agricultural landscape occur only in southern Sweden, their elegant flight and narrow wings reminded me of the long-tailed skuas I saw at Pielavalta.

The carpet of flowers is more vivid on Öland than on Sarek's hillsides. Öland's limestone plateau, or *alvar*, is topped by calcareous soil that in early summer is ablaze with orchids including Adam and Eve, military and burnt-tip.

A plant ecologist might be reluctant to label classic Öland specialties like hoary rock-rose or fern-leaf dropwort as calcareous flora, even though they grow exclusively on chalk and limestone soil. It is not the chalk or limestone that holds the key here, but the calcium ions that drive phosphorus from the soil solution when they form insoluble compounds with phosphate ions. Species like dock and speedwell, which avoid calcareous soil, quickly suffer the effects of phosphorus deficiency if they germinate in calcareous soil, in contrast to typical alvar species which have adapted to life in calcium-rich earth by being able to absorb phosphorus even when it is in short supply. Dock will, however, thrive in topsoil from which calcium has leached due to acidic rainwater, and in such conditions can pose a threat to other alvar-dependent species. Many calcareous sandy steppe areas in the eastern Skåne region of southernmost Sweden have suffered degradation of this sort, though efforts to remove the topsoil or restore calcium-rich soil to the upper layer have successfully benefited the calcareous grassland flora.

Role of Phosphorus Deficiency and Mycorrhizal Fungi in Floral Blooms

The rich flora of the Öland alvar and Sarek's magnesite quarry epitomises the resourcefulness of plants in adapting to phosphorus scarcity. Some species secrete organic acids from their roots that enable them to absorb phosphate ions. Many enter a symbiosis with mycorrhizal fungi to augment their phosphorus intake in return for providing access to the carbohydrates the plants produce during photosynthesis. Some plants even specialise in parasitising others' adaptations in order to gain access to phosphorus. Inadequate supplies of phosphorus limit the growth of all plants but also create scope for more life forms than a nutrient-rich environment where a handful of fast-growing species can outcompete the others.

Most Scandinavian plants are colonists from calcium-rich areas in central Europe, from where they spread slowly while gradually evolving into different species. Acidic soils are less common in central Europe than in more northerly latitudes, which is why fewer acidic-soil species have evolved in Scandinavia. The reverse is true in Australia, where calcareous soil is uncommon and therefore plays host to fewer species than low-calcium soil. Even the most attractive soils from a flora perspective have a lack of phosphorus owing to high concentrations of aluminium and iron that fix phosphate ions. Many plants here grow hairy roots that secrete citric acid, which enhances phosphorus availability by dissolving aluminium and iron phosphate. The strategy has not developed to an equivalent degree in Europe's calcareous soils because the change in pH is deleterious to mycorrhizal fungi, which are a more important source of plant phosphorus.

The symbiosis between mycorrhizal fungi and plants was discovered in the 1930s in Florida during successful attempts to control a parasitic fungus that was attacking the roots of orange and lemon trees. Scientists found that the saplings would not grow, even when the fungus was removed, unless large quantities of phosphorus were added. On closer inspection, they discovered another type of fungal structures on the wild saplings that were absent in their cultivated congeners. The scientists suspected this fungus was helping the wild saplings to absorb phosphorus—a hypothesis they proved by adding soil from wild orange and lemon trees, which allowed the fungus to recolonise and the cultivated saplings to grow normally again. Since then, Florida's orange and lemon growers always ensure their trees' roots are colonised by mycorrhizal fungi.

Scientists in the Netherlands have investigated the importance of the role played by mycorrhizal fungi for plant structure in different places and environments. They have been able to demonstrate that if you sterilise earth from a flower meadow, put it in pots and plant it with meadow flower seeds, it is not possible to replicate the meadow biotope. All that will grow are a variety of grass species; the flowers remain conspicuous by their absence. Things change radically, however, if you add spores of mycorrhizal fungi. All of a sudden, the

Phosphorus provider. When *Carex* and other sedges can grow amazingly intricate root hairs when exposed to phosphorus deficiency. These produce organic acids that lower the soil pH to release the elusive phosphorus. The root hairs only grow in extremely phosphorus-poor soils and are absent in nutrient-rich farmland. But if we could breed sedge varieties to grow root hairs even in nutrient-rich conditions, we could perhaps access the phosphorus that lies trapped in farmland from decades of fertiliser use, thereby tapping a valuable resource at a time when global phosphorus reserves are running low

meadow flowers germinate and bloom. This is because many flowers have thick roots that need mycorrhizal fungi to absorb phosphorus. If the fungi are absent, the flowers are outcompeted by grasses, which have more delicate root systems better able to absorb phosphorus unaided.

Mycorrhizal fungi are not alone in affecting plant life. Other fungi can infect plant roots and even kill the host. Rare plant species are often especially sensitive to such fungi, their very rarity being the result of having to colonise new areas to mitigate against fungal attacks. Commoner species have generally evolved more effective defence mechanisms. Some rarer species manage to retain a foothold by

Competitive grass. Grass has a finely branched root system reminiscent of a mycorrhizal fungus mycelium. The vast majority of plant species have less developed root systems and depend more on collaboration with fungi to ensure nutrient uptake. Fertilising nutrient-poor land benefits grasses but disadvantages species that depend on mycorrhizal fungi. Adding fertiliser to a meadow full of wild flowers can rapidly turn it into monotonous grassland

choosing areas subject to disturbance. Marram grass, for example, grows in sand dunes that are continually being shaped and shifted by the wind. Their roots are sensitive to fungal attack but the plant is periodically covered in sand, which enables new roots to grow and replace the old ones as they perish. If the sand dunes stabilise, resulting in the sand remaining fixed in place, marram grass cannot survive for long.

Some Plants Thrive in Acidic Soil

Some plants thrive in acidic soil. Growers of American blueberries know that you have to mix generous amounts of unfertilised peat into the soil or else the leaves bleach due to insufficient nourishment. Iron deficiency is a particular problem for blueberry bushes because iron becomes more soluble in acidic conditions and harder for the bushes to absorb when the soil's pH rises. Some species have evolved to manage this problem by secreting molecules that bind iron ions for enhanced absorbability. In strongly acidic soil, plant roots often suffer from increased concentrations of toxic aluminium ions. Plant species that can tolerate

Peat-loving rhododendrons. If you want to plant rhododendron, or other heathers, in your garden, it is best to replace the ordinary soil with coarsely shredded unlimed peat. These plants thrive in acidic soil and form a symbiosis with ericoid mycorrhizal fungi, which are experts at releasing nitrogen-rich amino acids from peat and other organic material. Weeds require easily accessible nitrogen and are rarely a problem in these conditions

highly acidic soil do so by protecting their roots through the secretion of substances that bind the aluminium ions.

Crowberry, blueberry, bearberry and other members of the heather family predominate in tundra environments, where the soil is acidic and nutrient-poor. They have adapted to this environment by evolving thin, hairy roots that form dense lattice networks in the organic topsoil. Cells in these hairy roots are full of fungal hyphae of another mycorrhizal type than that found in grass and herbs. The mycorrhizal fungi found in members of the heather family secrete enzymes that release nitrogen and phosphorus from complex organic compounds in the soil.

A few years ago I took part in an experiment in Abisko in which we added radioactive carbon dioxide to tundra plants enclosed under plastic covers. The plants absorbed the carbon dioxide through photosynthesis and the radioactive isotope enabled us to observe the carbon on its journey down into the soil. I asked one of my students to extract roots from the soil so we could analyse how much carbon entered the fungal tissue that filled the cells, a task that proved virtually impossible as the samples were enmeshed in spidery roots that were extremely difficult to separate from the surrounding soil. Nevertheless, after many painstaking hours under the microscope, we eventually developed enough samples and were able to see that the fungi utilised the plant's photosynthetic products heavily during the major part of the growing season.

Rich Flower Meadows Rely on Man's Management

Geology and fungal presence are not the only factors that determine what plant species can grow in which soil: the way we humans manage the land is also critical. Traditional hay meadows and pasture are rich in flora but difficult to replicate in a suburban garden, though you might suppose otherwise considering all the meadow seed varieties you can buy at your local garden centre. Garden soil is often too high in nutrients, and depleting soil of nutrients is no easy task. My father would sow meadow flower seeds on a large patch of land at his summer cabin in southern Sweden. But even when the seeds germinated, it did not take long for grass to take over and outcompete the flowers. A few years ago he began mowing the meadow with a scythe and removing the cuttings. The results have been promising and slowly the land is starting to look like a meadow. Removing nutrients from the soil is a painstaking job, but worth the effort in the knowledge that a meadow rich in flowers will quickly be ruined by adding fertiliser.

At Hörjelgården Farm in eastern Skåne, southernmost Sweden, the land has been carefully managed in recent years to restore it to how it would have looked in the eighteenth century. Schoolchildren visit the farm to learn about how people farmed and lived on the land in those days. The grass meadow is cut with a scythe and the hay used to feed the farm animals during winter.

Back in the old days, the tradition was to cut the meadow when the seeds pods of the greater yellow-rattle began to rattle in the wind. Only then was the hay dry enough to be stored for winter (eighteenth century farmers had no fans for hay-drying). The flowers that seeded before the meadow was cut survived until the following year; late-blooming species had to look elsewhere for a foothold.

The big challenge for eighteenth century farmers was to avoid depleting the soil. Meadows provided winter fodder for livestock, whose manure was used on the fields to add nitrogen and phosphorus. This process depleted the meadows, but typical meadow flowers, with their small leaves that are not energy-intensive to grow and slow growth cycles, are highly economical with nutrients. A diverse array of species can share the space when no single species grows large and dominates.

Nutrient-intensive species like nettles and broad-leaved grasses survived on compost heaps and near outdoor toilets by growing extensive root systems, but were unable to survive on the meadows. Nutrient-rich areas were uncommon in the agricultural landscapes of yesteryear, and most plant species were adapted to nutrient-poor conditions. Today, nutrient-requiring species have expanded and the nitrogen content of the soil has increased because of man-made pollution.

The big shift coincided with the introduction of artificial fertiliser in the first half of the twentieth century. Until then, nitrogen-fixing bacteria held a monopoly on converting atmospheric nitrogen into amino acids and proteins. Then came the Haber–Bosch process, one of the great innovations of the modern age and a breakthrough that earned its authors, Fritz Haber and Carl Bosch, the Nobel Prize in Chemistry in 1918 and 1931 respectively. The nitrogen produced in their process was mixed with phosphorus mined in North Africa to

Artificially fertilised algal bloom. Nitrogen supply has increased in the industrialised world during the last century. Nitrogen oxides are formed during combustion and are released by car engines, for example. Ammonia—a compound of nitrogen and hydrogen—evaporates from livestock barns and manure piles. Grass grows better, but at the expense of flowering plants, when this nitrogen reaches the soil. Algae also benefit, forming a near-phosphorescent coating on tree trunks, buildings and, as here, on fence posts

create fertiliser capable of exponentially increasing harvest volumes and rendering barren land fertile. Haber and Bosch were actually not the first to manufacture artificial fertiliser: that distinction falls to Kristian Birkeland, a Norwegian better known for explaining the nature of the *Aurora borealis*. Most spectacularly, he also invented an electromagnetic cannon with a range far exceeding the howitzers of the day and which, Birkeland claimed, was also silent when fired. When Birkeland demonstrated this wizardry at the Oslo University banquet hall in 1903, an almighty roar filled the air when he turned on the electricity. The accompanying flash of light sent the audience scattering in panic. Birkeland himself was delighted, calling it the most dramatic moment in his life and later remarking that "the projectile shot down my shares from a value of three hundred kroner to zero, but hit its target perfectly". This was because he had realised that atmospheric nitrogen reacted with oxygen during the release of energy from the flash of light. It was well known by then that nitrate was created in this way when a lightning bolt hit the ground, and Birkeland realised his cannon could be developed into an oven to make artificial fertiliser. Such a venture would be energy-intensive so Birkeland teamed up with an industrial magnate who bought one of Norway's waterfalls, Rjukanfallet, to build a hydropower dam. During the course of twelve months they manufactured a couple of kilograms of artificial fertiliser. Of course, this did not get them very far and the oven was a commercial failure. Not until the Haber–Bosch process was invented did it become possible to manufacture artificial fertiliser on an industrial scale.

Hoofmarks. Livestock grazing and trampling shapes plant communities. Grass withstands grazing because the animals cannot reach all the way down to the leaf buds at the base of the plant. The buttercup forms substances that cows cannot tolerate and grows commonly in grazed fields. Grazing also benefits low-growing species that would otherwise be overshadowed by tall grass. Some species thrive in the disturbed soils trampled by cattle, while others grow unhindered in thorny thickets the cows cannot reach

The advent of artificial fertiliser was a disaster for meadow flora, which were choked by grass when fertiliser was spread on meadows. Much of the nitrogen ends up in watercourses and ultimately in the sea, where it can be a major hazard. Increased concentrations of nitrogen and phosphorus in marine environments can result in algal blooms. Eutrophication is today one of our most serious environmental problems, on a par with the twin challenges of climate change and the mass extinction of species that are accompanying our unrelenting exploitation of the Earth's resources.

On pastureland, grazing by animals is the key factor that dictates the local flora. It is important for plants in this environment to preserve their growth centres sufficiently deep below the surface to stay out of the reach of grazing livestock. Likewise, thistles grow prickles to deter grazing animals, while some plants exploit the cover afforded by blackberry scrub as a nursery. Other species, such as buttercups, emit poison to dissuade animals from eating them. Fortunately, buttercup toxins are destroyed when the leaves dry, making hay safe to eat. Some flowers will choose to bloom beside heat-absorbing rocks, while others prefer a damp hollow.

The pasture and meadow at Hörjelgården farm has never been cultivated or fertilised. Here, up to forty different species of plant can be found in a single square

Thoughtful thorns. In Africa, the leaves of the acacia tree are attractive food for grazing animals such as wild gazelles, sheep and goats. To avoid being grazed bare, the tree has evolved strong thorns on its lower branches within animals' reach. Higher up, towards the crown, it does not need to waste its resources on these costly constructions

◀ *Solitary bees.* Adjacent to my plot of land in Torna Hällestad, southern Sweden, lies a field known as "rabbitland". In the mornings and evenings, the sunlight plays on the dry grass, making it almost look as though it were on fire. Local council workers regularly rough up the field with machinery to create optimal habitat for wild bees, which thrive in sand exposed to the sun. Controlled burning is also carried out to reduce the spread of tall oatgrass. On this morning, frost is sparkling in the remains of the burnt grass

Not all bees form colonies; some live alone. The open sand at rabbitland creates favourable conditions for solitary sand bees (*Andrena vaga*), which excavate burrows in the sand. Here a female sand bee looks out of her nest. The area is protected to provide a sanctuary for the endangered bees

metre, including rare grasses such as mat-weed, common heath grass and quaking-grass, and flowers ranging from orchids to mountain bugle. This biotope has suffered a drastic decline since the eighteenth century and only remnants survive nowadays. Even if we seek to preserve the last vestiges by creating nature reserves and banning fertiliser use, these areas remain vulnerable to human impact in the form of nitrates carried in rainwater and emanating from car exhaust fumes as well as ammonia released into the air from livestock farming.

How Child's Play Helped Flowers and Bees

Just as decades of careful management have led to the creation of agricultural zones rich in species that conservationists are at pains to preserve, so human disturbance has in some cases inadvertently created valuable biotopes. Near my home in Torna Hällestad, southern Sweden, is a large sandy expanse known locally as "rabbitland". I sometimes go there on bright summer nights, slipping through a gap in the hedge to enjoy the magical glow of evening primrose. The petals of this bright yellow flower hang slackly during daylight hours but glow luminously at night to attract night-flying insects. A boulder-ridge, topped by a tree-fringed farm, and adjacent potato fields used to overlook the area but this land was excavated in the 1950s and 1960s to extract gravel from the ridge for road building. When gravel extraction stopped, drought-tolerant plants began colonising the sandy soil and nature slowly recovered its foothold. Besides evening primrose, the area now hosts dwarf everlast, viper's bugloss, goldmoss stonecrop and large clumps of sheep's bit.

The sandy ground became a haven for rabbits and my neighbour's dachshund honed a special technique for opening the back door so it could go out and hunt the bunnies. The locals also enjoyed the influx and people still talk of garden cages baited with carrots. Once seven or eight rabbits had entered a cage, the home-owner would heave it towards the house using a rope. If they were lucky, a couple of rabbits might still be in the cage—enough for a nice stew. A wave of myxomatosis in the 1970s decimated the rabbit population, and the species is now scarce here. Conservationists want them back in numbers to control the false oat-grass that takes over if nitrogen content becomes too high in the soil. False oat-grass outcompetes low-growing species like dwarf everlast and stonecrop. Rabbits can only graze tender false oat-grass shoots; controlled burning is needed when the grass gets too thick—a method that also combusts some of the nitrogen content. Another method is to grow rye and then harvest the crop as a way of removing soil nutrients.

Jasione and stonecrop are popular sources of pollen and nectar for the many insects that live at rabbitland—and the combination of a rich flora and sandy soil makes the area one of the best in Sweden for wild ground-dwelling bees. Forty-five different species occur here, many of them threatened and in decline both in Sweden and internationally.

In the middle of the area is a gentle hollow where I often pause on my nightly strolls. We used to call it the long-jump pit when the children were small, and we ran up and down setting new records as the kids grew up. The hollow, dug by local children using buckets and spades in the 1970s, is today one of the best spots for the ground-dwelling bees that have brought the area to the attention of entomologists. The loose sand that slides down the sides of the sandpit creates perfect conditions for the bees. The local authority has copied the sandpit concept by excavating similar hollows throughout the field to improve conditions for the bees. Perhaps it is due to the children that bees now thrive here and large sums of money are being invested to keep encroaching vegetation at bay.

Nature Can Self-pollute

Local changes in soil chemistry can sometimes cause abrupt changes in the local flora. Heavy metals occur naturally in some soils, and just as they are toxic to humans, so they can be harmful to plants in high concentrations, though some species have learnt to adapt to such conditions.

Plant mapping is used widely internationally to discover new mineral deposits. Geotechnical engineers discovered the area in Kastad, Norway, with high concentrations of lead in the soil (see previous chapter) by studying changes in the flora using satellite images. Alpine catchfly has been used to indicate copper deposits and alpine chickweed to denote the presence of serpentine, a rare mineral created under special conditions during the formation of mountain ranges by movements in the Earth's tectonic plates. Serpentine presence results in nutrient-poor soil with high concentrations of heavy metals such as nickel, chromium and cobalt. Most plants—apart from alpine catchfly—cannot tolerate such environments. Various similar examples can be found around the world. In the US state of Oregon, for example, Douglas fir gives way to pine in areas of serpentine soil as the latter can tolerate the heavy metals present in the soil.

Serpentine soil is rich in magnesium but low in potassium, which often leads the plants growing there to suffer from a deficiency. There is usually enough magnesium in most soils, but sometimes the high magnesium content in serpentine soil can be critical for plant survival. When, in the 1970s and 1980s, spruce forests in

Heavy metals. This cross-section comes from a spruce that grew very slowly because it stood on serpentinite bedrock. Over twenty-five years, the trunk diameter increased by just two centimetres, whereas another nearby spruce, which stood on amphibolite bedrock, grew as much in four years. Soils on serpentinite bedrock are high in heavy metals but low in potassium, which is the most critical element for plants after nitrogen

then Czechoslovakia were dying due to acid rain caused by the burning of sulphur-containing oil and coal, some scientists feared Swedish forests would suffer the same fate. Acid rain leaches key nutrients such as calcium, potassium and magnesium from the topsoil, causing many Czechoslovakian spruce trees to die from a lack of magnesium, a necessary ingredient in chlorophyll and hence essential for photosynthesis. When I visited the country in the late 1980s I saw large areas of dead forest in exposed mountainous areas near the German border, while spruces had fared much better on magnesium-rich serpentine soil. In Sweden we are lucky because large quantities of magnesium are carried inland on wind from the sea. Also, we were not as heavily affected by sulphur emissions as former Czechoslovakia, where vast quantities of sulphurous lignite were burnt for energy production during that era.

Some plants are capable of storing large concentrations of heavy metals in special structures, effectively neutralising them. Such plants, such as willow—known for its ability to store cadmium—can be used to remediate contaminated ground. Pollution from cadmium compounds in artificial fertiliser is a widespread problem in agricultural soils. Willows grown near water treatment plants can effectively purify the water and be later harvested for biomass. It is, however, important to collect the ash and dispose of it safely to avoid re-releasing it to the environment.

Arsenic is a deadly heavy metal and one of the most carcinogenic substances known to science. It occurs naturally in certain types of rock and is responsible for polluting many wells in India and Bangladesh. However, a potential solution has emerged in the form of a Chinese fern that can store arsenic in its leaves. The fern will only grow in warm and moist climates but scientists hope they can use genetic engineering to graft its arsenic-storing properties onto other plants, especially those with large root systems capable of purifying substantial expanses of ground.

But soil contains much more than plant roots. It is teeming with microorganisms, and it is here that soil biodiversity reaches its zenith, as we will discover in the next chapter.

Further Readings

Billqvist M (red) (2002) Det skånska kulturlandskapet. Naturskyddsföreningen i Skåne
Craine J (2009) Resource strategies of wild plants. Princeton University Press, New Jersey
Grundsten C (2000) Sarek och Kebnekaise: där ljuset skiftar. Fernström Publishing, Sweden
Jago L (2002) The northern lights. Vintage, New York
Peterson LR et al (2004) Mycorrhizas. Anatomy and cell biology. NRC Research Press, Canada

Life in the Soil

The worms mingle the whole intimately together,
like a gardener who prepares
fine soil for his choicest plants.

Charles Darwin

Abstract The greatest biodiversity is in the soil, and this chapter describes the soil's inhabitants—everything from fungi and bacteria to single-celled organisms, mites and other bugs. How are they connected in the foow-webs of the soil? Who eats whom? Fungi and bacteria are important for the decomposition of the organic material in the soil, yet they work in different ways, which affects the circulation of nutrients. Tropical rainforests have a diversity of species above ground whereas in northern latitudes the diversity is to be found under the surface. Certain creatures, such as earthworms, affect the structure of the entire soil ecosystem and accelerate the circulation of nutrients. But when European earthworms invade forests in Minnesota, biodiversity suffers. Termites can eat wood thanks to symbiotic microorganisms in their digestive tract; as they gather material to feed their larvae, they affect the soil's fertility and the distribution of plants on the savannah. In addition, they have contributed to the cultural development of the Australian Aborigines.

Diversity of Species in the Soil

Soil teems with life. A single gram can contain several hundred metres of fungal hyphae and more than ten billion bacteria, along with microscopic worms like nematodes and unicellular organisms such as amoebae and protozoa. Larger organisms including springtails, mites, beetles and millipedes are present, too, but in much smaller numbers.

One might well wonder how many organisms a child ingests when it sticks dirty fingers in its mouth when playing outside. Probably billions. But eating a few crumbs of soil would not normally make a child ill as most microorganisms are perfectly harmless. There are, however, a few soil bacteria that one is advised to

H. Wallander, *Soil*, DOI 10.1007/978-3-319-08458-9_4

◀ *Nature's great larder.* Eating soil is a tradition in a wide variety of cultures. Do pregnant women in Africa eat loamy soil due to calcium or zinc deficiency—or because the clay neutralises toxins in the stomach and alleviates diarrhoea? Evidence suggests the nutritional value of certain types of food rises if they are eaten in combination with clay. Acorns are widely ignored as human food due to their high tannin content. But Native Americans in California and the indigenous population of Sardinia would mix acorn meal with clay because it neutralised the acorn tannins, rendering the bread they baked edible. In the West we have used clay as medicine. The highest-selling diarrhoea remedy in the USA—kaopectate—originally contained kaolin, an active ingredient that is a kind of clay

avoid. Tinned vegetable manufacturers take great care to wash all soil traces away to ensure no *Clostridium* bacteria enter the tin. If the contents are not heated for long enough, the bacteria may start reproducing in the anaerobic conditions inside the tin, giving off a nerve gas that causes botulism, a life-threatening disease. But *Clostridium* bacteria occur in soil in such small concentrations that they pose no risk to humans who inadvertently swallow a little soil.

Swallowing soil is not confined to children. In some cultures, soil is eaten for medicinal reasons, which may not be as misguided as it sounds. Modern science has shown that some plant toxins are neutralised if mixed with loamy soil. It is well known, for example, that the potato plant is poisonous. Yet the tubers we eat carry no poison if they grow in the dark so they do not go green. The indigenous wild potato in South America also had toxic tubers, which indigenous people neutralised by soaking them in loamy soil slurry (the toxins attached themselves to the loam particles). Pregnant women in some African tribes eat soil for its calcium, which they need for breast milk – a custom particularly prevalent among people who do not keep dairy animals. Many animal species eat soil, both as a source of salt and other minerals and to purge themselves of toxins absorbed from plant food.

Long known as an important reservoir of biodiversity, soil is difficult to study because many of its resident organisms are small and hard to detect. The standard way of measuring biological activity in soil is to analyse the volume of carbon dioxide produced by soil-dwelling organisms. Microscopic fungi and bacteria contribute the lion's share, emitting ten times more carbon dioxide than all other animals put together.

I once attended a lecture by a soil microbiologist who told the audience, "Imagine yourself as a birdwatcher, and the only birds you've seen so far are a hen, a goose, a budgerigar and a duck. Then someone takes you to a tropical rainforest and shows you the array of birdlife there, overwhelming you with all the brightly coloured parrots and the hummingbirds zipping around from flower to flower. That's what it's like for soil microbiologists. In the past, we were only able to study species you could cultivate in a Petri dish, which equates to a few percent of all the fungi and bacteria that are actually present in the soil. Using modern techniques, we can extract DNA from soil samples and divide the sequences into fragments that enable us to identify the organisms. We often discover unique DNA sequences from species new to science.

Take archaea, for example. They resemble bacteria and used only to be recorded in extreme environments like salt lagoons, hot springs or cows' stomachs,

where they convert carbon dioxide into methane. In recent times they have been found to be abundant in most soils, and as a result have been classified as a separate domain alongside the other two domains of life: bacteria and eukaryotes [organisms whose cells contain a nucleus]. Future research will doubtless reveal them to be more important than we could ever have imagined.

◀ *Diversity above and below ground.* Mention high biodiversity and many would think of the Amazonian rainforest and its thousands of tree species, hundreds of bird species and countless insect species. A single leaf in the rainforest can host lichens, algae and all sorts of bugs. Scandinavian coniferous forests, on the other hand, are poor in species, with few plants, animals and fungi above ground. Here, the odd birch is visible among the spruce trees, while lingonberry bushes and moss dominate on the ground. But it is in the humus-rich soils of this type that the greatest biodiversity is to be found, even though we have not yet managed to identify more than a fraction of all the species present

Advances in molecular biology have improved insight and understanding of this diversity. Soil microbiologists have acquired tools to study the most prevalent species in the soil—not only those that can be cultivated in a petri dish. The spruce forest in the photo contains around one hundred species of mycorrhizal fungi. Boletes and brittlegills appear as mushrooms in the autumn but live for many years as mycelia underground

The lecture insights were still with me when the Swedish Royal Academy of Sciences in Stockholm invited me to give a lecture of my own on biodiversity in Swedish spruce forests.

Thus I found myself standing in front of an assortment of scientific luminaries seated in comfortable armchairs. I began by showing images from the Amazonian rainforest, a familiar environment and home to hundreds and thousands of bird and insect species respectively. Then I asked the audience to don a pair of special glasses and join me on a trip below ground to explore diversity on another, much less familiar, scale. Interest visibly grew as the academy members watched my images of the rainforest's thin topsoil and near-sterile red earth lying directly beneath the fine upper roots of the trees and their sparse covering of leaf litter. The leaves were suffused with fungi, which along with bacteria decomposed the leaves and created a supply of nutrients for the trees. Few other organisms were present in the thin soil.

We then shifted to a Swedish spruce forest, where thick blueberry scrub and a lush layer of stair-step moss covered the ground. Amid the spruce stood a few birches and aspens but nothing like the diverse array of tree species—thousands in a single hectare—found in the Amazon. By contrast, the thick soil is home to a host of fungi and bacteria—a diversity that the barren rainforest earth cannot come close to matching. Carefully pull up the turf surrounding a boulder and you will find a complex network of roots and hyphae. The hyphae belong to mycorrhizal fungi that live in symbiosis with the trees. They extend the surface area of the tree roots many times over, absorbing and transporting water and minerals to the roots in return for access to carbohydrates synthesised by the tree. This symbiosis is different from the mycorrhizae that developed with the first terrestrial plants, as described in the "How is soil formed?" chapter. A far greater number of mycorrhizal species exist in symbiosis with trees. Indeed, most of the fungi we see on a typical walk through the forest—milk caps, boletes and chanterelles, to name some of the commoner groups—coexist closely with trees. Some, like the birch bolete, are restricted to a specific species; others, like the golden chanterelle, can form mycorrhizal partnerships with different trees. A single forest can host hundreds of different species.

Worldwide there are several thousand species of fungi that form mycorrhizal partnerships with plants, but the number pales in comparison to microfauna like nematodes, of which there are more than thirty thousand species, and unicellular

◀ *Filament lifelines.* Fungal mycelia from the roots of a spruce penetrate through the ground. Many boletes form white mycelia similar to those in the photo, but the yellow mycelium belongs to a fungus that never forms any fruit bodies. A pine needle is visible for comparison in the upper part of the image and a spruce root appears to the lower right

The king bolete in the bottom photo forms mycorrhizae with both deciduous and coniferous trees. It is recognisable by the white reticulations on its stem. Intact specimens of this mushroom are among the most prized by pickers

organisms, of which over sixty thousand have been described. The true numbers are even bigger. One of the pioneers of nematode research wrote a century ago that one could remove everything except nematodes from the ecosystem and be unable to notice any immediate impact: mountains, valleys and flora would look much the same. So numerous are nematodes that the numbers are almost mind-boggling. A single square metre of soil can hold up to ten million of them.

After my lecture to the Royal Academy of Sciences I visited the adjoining Bergius Botanic Garden. Strolling round in the hothouses, I wondered if my message had got through. A single spadeful of earth from a Swedish forest contains more species of microbe than all the plant and animal species found in the world's tropical rainforests.

The Soil's Web of Life

A question that soil scientists have long struggled with is how all these life forms interact. Who eats what? In a forest or savannah grassland it is relatively easy to see which animals are herbivores and which are carnivores. Plants serve as food for gazelles on the savannah or leaf-eating larvae in the forest. In turn, these animals become food for predators like lions or insectivorous birds. Sometimes there is an extra link in the chain: small birds are food for birds of prey, for example. But what does the food chain look like below ground?

One starting point is the organic exudates secreted by plant roots. In part, these secretions serve as a lubricant for the roots as they grow through the soil and as food for mycorrhizal fungi. In many plants, the soil adjacent to the roots hosts a rich flora of root bacteria that thrive on this ready source of carbohydrate. Root bacteria can reproduce one hundred times faster than bacteria living a millimetre or so away from the root surface. The carbohydrate glut also attracts bacteria-eating organisms like amoebae, protozoa and nematodes, whose presence in turn benefits plants because nitrogen in the bacteria is released during the feeding process. Some bacteria secrete substances that stimulate root formation, a beneficial role that tends to result in predatory organisms ignoring rather than eating them. Aphids can multiply when soil organisms facilitate plant growth, and hence plants serve as a link between the fauna above and below ground. Scientific studies indicate that the interaction between these different worlds is far more complex than previously thought.

Another starting point for the soil food chain is the organic matter present in the soil, originating from dead leaves, roots and insects and other organisms.

Breathing space. There is no doubt that human health benefits from regular opportunities to spend time in the forest. Perhaps the light filtering through the foliage makes us feel calm and harmonious inside. This may be what Antonio Gaudí had in mind when he designed the Sagrada Familia cathedral in Barcelona. The pillars supporting the church branch out like trees, and the light streaming in through windows in the ceiling makes you feel as if you were in the middle of a wood

Fungi and bacteria are quickly on the scene when something dies. The latter thrive in nutrient-rich environments and when worms and other digging animals disturb organic matter. Fungi, on the other hand, favour nutrient-poor conditions and prefer peace and quiet because the action of worms damages their delicate hyphae. If left undisturbed, fungi can develop large hyphal networks capable of absorbing carbohydrates in one place and nitrogen in another. Unlike bacteria, they do not necessarily rely on all nutrients being present in one spot.

Some fungi have developed a capacity to apportion their resources over a very large surface area. Sick building syndrome is often caused by mould growing in damp wood, but it is *Serpula lacrymans*, otherwise known as dry rot, that is every homeowner's worst nightmare. It can attack dry wood provided it has access to moisture from the ground outside the house. *Serpula lacrymans* transports this moisture through its tissue, which grows in thick masses along the stone foundations at the base of the wooden structure. As a result, dry rot can grow even on dry wood. Many species of fungi work in a similar way in the wild, overcoming dry patches of ground by transporting water through extensive networks of hyphae permeating the soil.

Soil's heterogeneous nature enables it to host a myriad of species. In a growing root, small organisms congregate round the tip, which is where carbohydrate secretion is greatest. But this changes as the root gets older, and other species gradually enter the scene as it grows. Conditions differ depending on the location in the soil profile. The fresh leaf litter lying on the floor of a spruce forest is high in carbon but low in nitrogen, making it a prime food resource for fungi that require carbon to build their biomass. Carbon dioxide is released as the leaf litter decomposes, and as the carbon content reduces so the nitrogen content increases. Mycorrhizal fungi operate a little deeper, where the nitrogen content is sufficiently high. Nitrogen is the most important tree nutrient; mycorrhizal fungi absorb it from the humus and exchange it for carbohydrates that the tree creates through photosynthesis. Farther down, other fungi absorb phosphorus and other mineral nutrients released during the weathering process. Here, oxygen and humidity play a role in determining which species are present. Some thrive in the airy passageways left by worms or rotting roots, while others prefer damper soil.

Nitrogen availability is pivotal to fungal diversity. Some mycorrhizal fungi absorb nitrogen through amino acids; others wait until fungi and bacteria have released ammonium and nitrates during the decomposition of organic matter. Nitrogen-poor forests are home to many species whose enzyme systems specialise in releasing nitrogen from humus, but in nitrogen-rich forests these are an unnecessary cost and the few mycorrhizal fungi that grow there invest in rapid growth instead.

Mushroom pickers looking for chanterelles do not like it when soil nitrogen content gets too high as this tasty fungus grows poorly in such areas. The chanterelle is in decline in nitrogen-rich areas across Europe, especially in the Netherlands, where nitrification is high due to intensive livestock farming. Nitrogen-rich ammonia from farm manure rises into the air and precipitates on nearby land.

Numerous small organisms eat soil bacteria and these in turn become food for other animals such as predatory mites and beetles. The various food chains in the soil mesh into a single food web in which animals like earthworms and termites play an especially critical role by turning over and changing the soil infrastructure. They eat decaying matter and let intestinal microorganisms work in their gut. The decomposition process accelerates when the decaying matter breaks down into small particles, allowing bacteria to work undisturbed and worms to benefit from the nutrients released by the bacteria.

Earthworms on a Lawn in Tomelilla

I expect to encounter earthworms when digging in my vegetable patch but sometimes come across them in less familiar places, as for instance when I attended a dance class at Tomelilla College of Further Education near my home in southern Sweden.

"You've got to move your body up and down in time to the music," explained teacher Max Somah, from Guinea in West Africa, on the lawn outside the school.

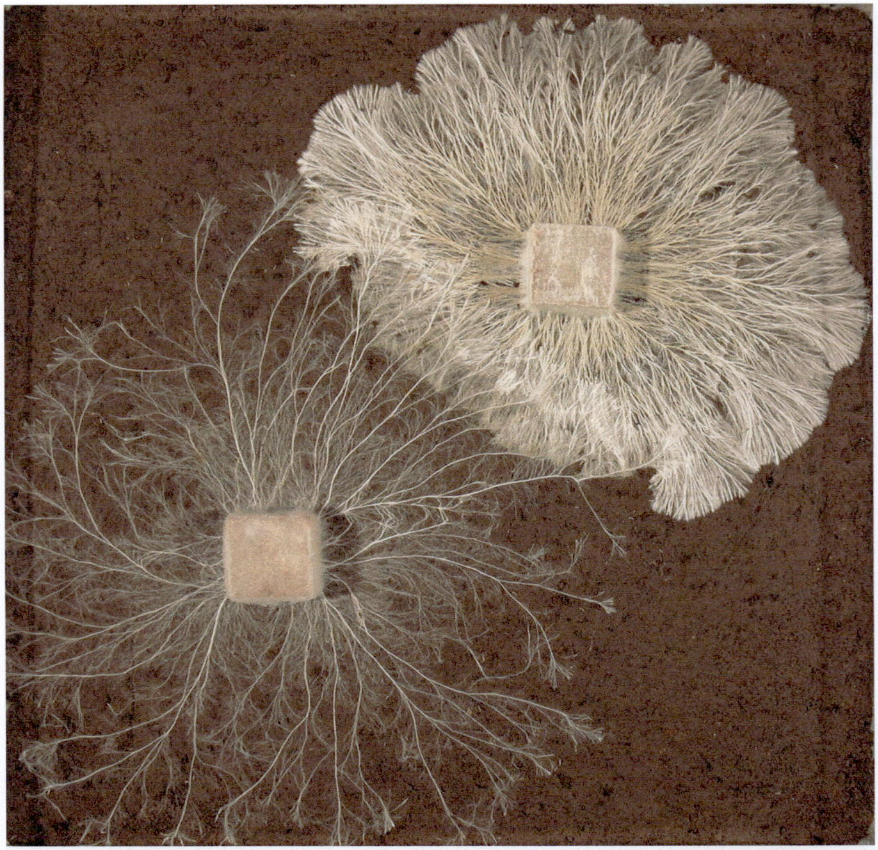

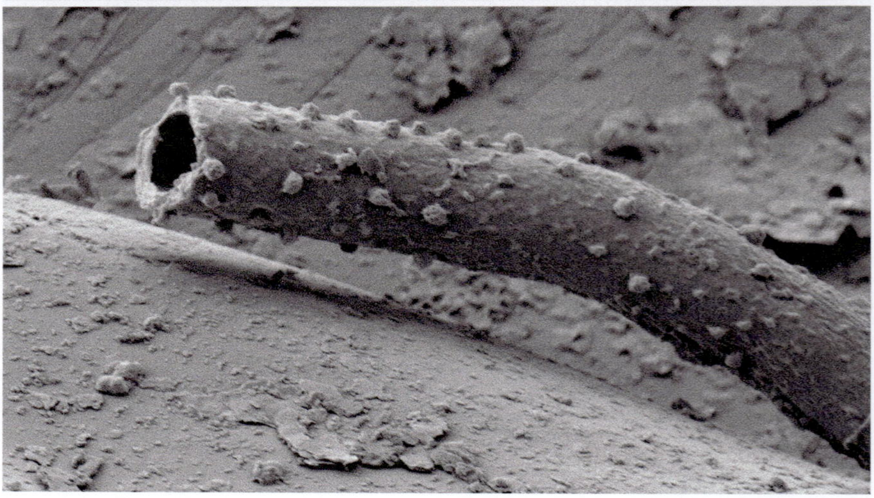

◀ *Organised foragers.* The fungi in the photo break down blocks of wood to gain access to energy. Wood contains almost no nitrogen, so the fungi grow out of the blocks with their mycelia to fetch nitrogen from elsewhere. The ability to source energy and nitrogen from different locations and then transport these resources within the mycelium makes fungi superior to bacteria in soil with uneven nutrient distribution

Fungal hyphae (*lower image*) are mostly hollow. The cell fluid lies in a thin film against the walls. It is only in the tips, where the hyphae are most active, that the cells are completely full of fluid. Many fungal hyphae have a granular surface due either to crystals formed by mineral residues that the fungi have dissolved to access nutrients such as phosphorus and potassium, or crystals formed to protect the hyphae from being eaten

After a while I began to grasp what he meant and gradually got into the music, my footsteps and arms moving in time with the drums as my stiff, crane-like movements gradually morphed into something suppler and more percussive. The drumbeats pounded through my belly and into my arms and legs.

Suddenly I slipped over and felt something sticky underfoot: a squashed earthworm. My fellow dancers were also slipping over. I tried to save the worms by throwing them away from the dance area so they could burrow back into the soil where they belonged, and it occurred to me that the dancing feet had tricked the worms into thinking it was pouring with rain. If water floods the soil's pores and hollows, earthworms must tunnel to the surface to avoid drowning. They must have been surprised to be greeted by hot sunshine and hopping human feet.

We regard earthworms as friendly creatures because they are so crucial to soil fertility, and we are happy when we see them while digging in the garden and sad if we accidently cut one in half. They are called angels of the soil in China for their ability to aerate the soil. In Ancient Greece, Aristotle likened them to intestines because the soil passes through them as they eat their way through it. Charles Darwin, father of the Theory of Evolution, spent much time studying earthworms and observed in 1837 that marlstone scattered on the fields as a soil conditioner ended up several centimetres below ground within a few years. This, he suggested, was the work of earthworms, though French scientists disputed the idea. Darwin continued to study earthworms, especially during his later years, and in 1881—six months before his death—he published his work on the species and its importance to soil formation. The book was a bestseller of its day, as popular as *The Origin of Species*. One reason was Darwin's conclusion that earthworms were intelligent creatures, a view at odds with the scientific consensus of the day. Worms were seen as slimy, repulsive pests that ate seeds and ruined the soil for growers. Horticultural manuals of the day described techniques for culling them by beating poles on the ground to entice them to the surface. People were encouraged to go out at night with lanterns and to despatch them in the same way we do today with Spanish slugs.

Darwin noted that earthworms were active at night and observed how they dragged leaves and other matter with them into their underground tunnels. He credited them with intelligence after watching how they pulled leaves with the sharp side first. Scientists today would disagree with Darwin that this behaviour indicated intelligence but would doubtless concur that it is fascinating how the animals ease their workload in this way. Darwin also advanced different theories on

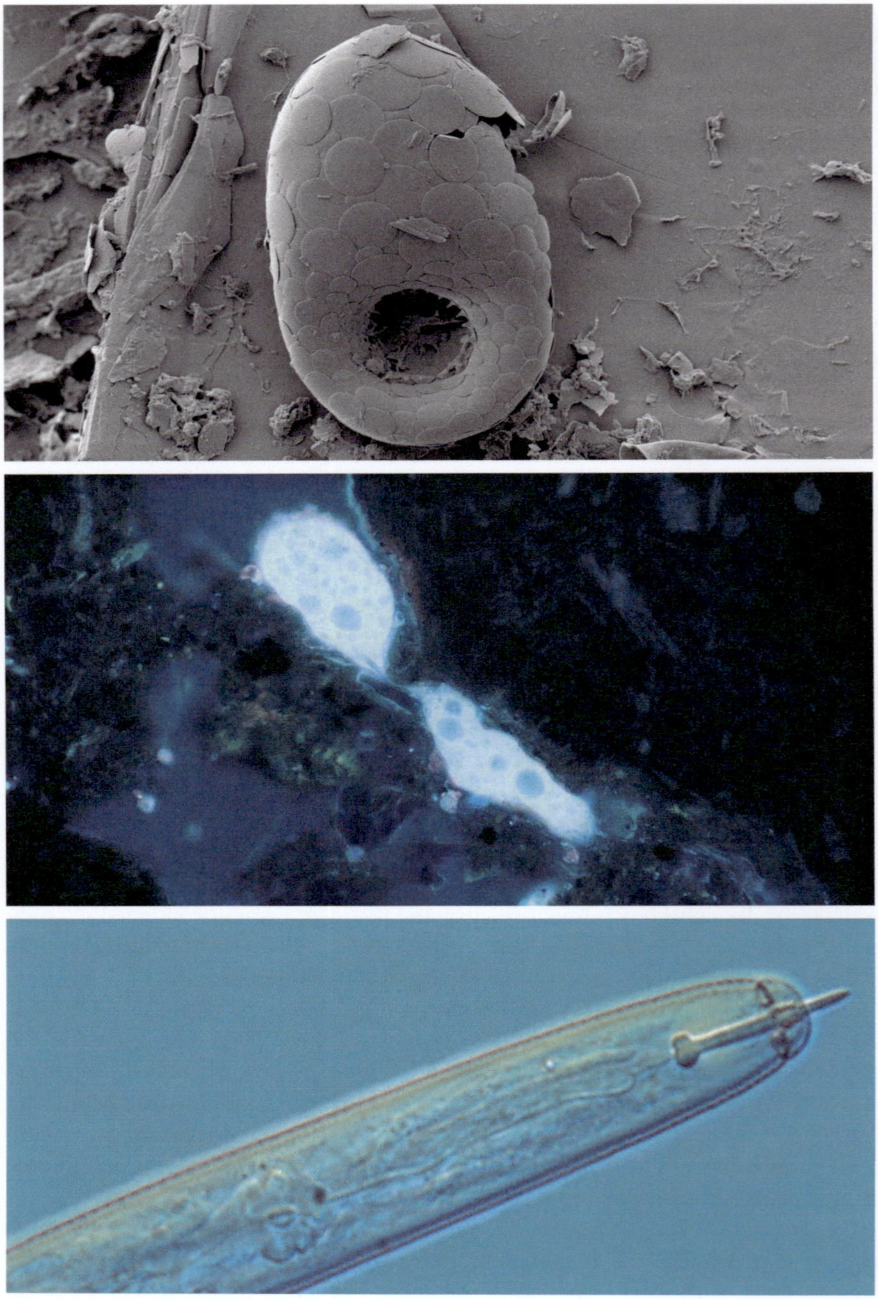

◀ *Inventive microfauna.* Amoebae are single-cell organisms that can number tens of thousands in a single teaspoonful of soil. Some have a shell, like the one in the upper image; others have no particular form but develop arm-like extensions that can capture bacteria, their principal food. In the middle photo, an amoeba moves within a pore and elongates its body when the pore becomes narrow. The amoeba is visible in white and the pore cavity is blue. A special group of amoebae burrow like vampires into fungal hyphae and suck out the cell fluid. They even go by the Latin generic name *Vampyrellids*

The most common multicellular land animals in the soil are nematodes, a sort of millimetre-length worm. Together with amoebae, they accelerate nutrient turnover when consuming bacteria and fungal hyphae. Some can cause disease: many of them in plants, others in animals. The plant parasitic nematodes in the bottom image puncture plant cells with their stylet and ingest the content

why earthworms took leaves into the soil, suggesting that they used them to insulate their chambers against draughts. Presumably he based this on Britain's damp and poorly insulated buildings.

Darwin also took an interest in how worms reacted to stimuli. One experiment involved placing worm-containing pots in his living room while his wife Emma played Chopin on the piano. The worms did not react to the music, but when Darwin placed the pots on the piano the vibrations caused them to tunnel into the soil. This reminds me of the lawn in Tomelilla and how the vibrations from African drums echoed through my body. Perhaps the worms were reacting to the same good vibrations.

By modifying the physical structure of the soil, worms facilitate decomposition of organic matter and accelerate plants' access to vital nutrients. It could take a hundred years for the earthworms to work their way through the soil of the Tomelilla lawn, during which they could produce up to one hundred tonnes of excrement per hectare in the form of the "castings" one sees on the surface. This may sound a lot but is nothing compared to what earthworms can achieve in the most fertile soils. The Nile river valleys are rich in earthworms and scientists there have estimated that they eat their way through two and a half thousand tonnes of earth per year and per hectare. Worm castings consist of organic matter, mineral soil and mucus from the worms' gut and the mucus-forming bacteria that live there. Darwin explained how mineral particles ground down the organic matter like a miller's stone. Worms live on the nutrients released by bacteria in their gut—a moist environment in which the bacteria thrive when breaking down cellulose and other compounds.

Earthworms' tunnels create pores that aerate the soil, allowing rainwater to penetrate the ground rather than collect on the surface, where it can damage the soil structure. Plant roots are known to be able to penetrate earth packed thickly by heavy farm machinery by utilising tunnels made by worms. I am always mindful of keeping my vegetable patch well aerated to create optimal growing conditions, digging nutrient-rich compost earth full of earthworms a good twenty centimetres down. Woe betide anyone who treads on the flowerbeds and compacts the soil: even with a spade or pitchfork you cannot recreate the earthworms' delicate tunnels.

In natural ecosystems, earthworms favour nutrient-rich soil in deciduous forest with a neutral pH. Here they carry fallen leaves with them below ground, where

◀ *Spectacular bugs*. Mites and springtails are the most common soil arthropods. Under the microscope they look like strange prehistoric animals. The springtail (*upper left*) takes its name from its long abdominal furcula, which allows the animal to leap into the air when threatened. Mites are among the most diverse animal groups in the soil. Some are predators: the yellow mite with the large front legs (top right) catches prey including springtails. Many species eat fungal mycelia, while others survive on dead organic material. The small yellow mites that have attached themselves to the legs of the large beetle larva are so-called lift specialists. The larva is not shown to scale

they digest them. This is why one does not see a thick layer of organic material in fertile broadleaf woodland in the way one does in coniferous and nutrient-poor deciduous forest. Deciduous forest with large earthworm populations is characterised by a rich understorey that may consist of high-nutrient-requiring species such as dog's mercury, goutweed and yellow wood anemone. Nutrient-poor broadleaf woodland, on the other hand, is often marked by a thick layer of dead leaves and no understorey, with the slow pace of decomposition partly reflecting the absence of earthworms.

Earthworms—Heroes in Europe, Aliens in Minnesota

Rural Minnesota in the United States was a favourite location for nineteenth-century Swedish settlers. Until then, the Minnesotan countryside there was earthworm-free but at some point earthworms were released to the environment, perhaps via imported plants. The effect was devastating. Minnesota's deciduous woods traditionally have a thick layer of organic matter aerated by beetles, millipedes and small rodents. The actions of earthworms in removing leaves from the surface deprives these creatures of their natural environment, and the plants that have adapted to rely on a porous layer of humus cannot grow in the hard, mineral soil. The result is lost biodiversity and recreational value. It may seem strange that earthworms can have such a diametrically different impact on different sides of the Atlantic, but this is due to plant evolution, which in turn is dictated by the conditions that apply in each specific spot. Environmental changes often deplete biodiversity, especially if the area has seen relatively little change for a long period of time, as is the case of tropical rainforests.

Termites as Engineers

Earthworms are often regarded as the engineers of the earth due to the way their tunnelling changes the soil structure. In the Tropics, termites perform an equivalent role by transporting large amounts of organic matter to feed their larvae. Termites exist almost exclusively on wood and one might well wonder how they

(continued)

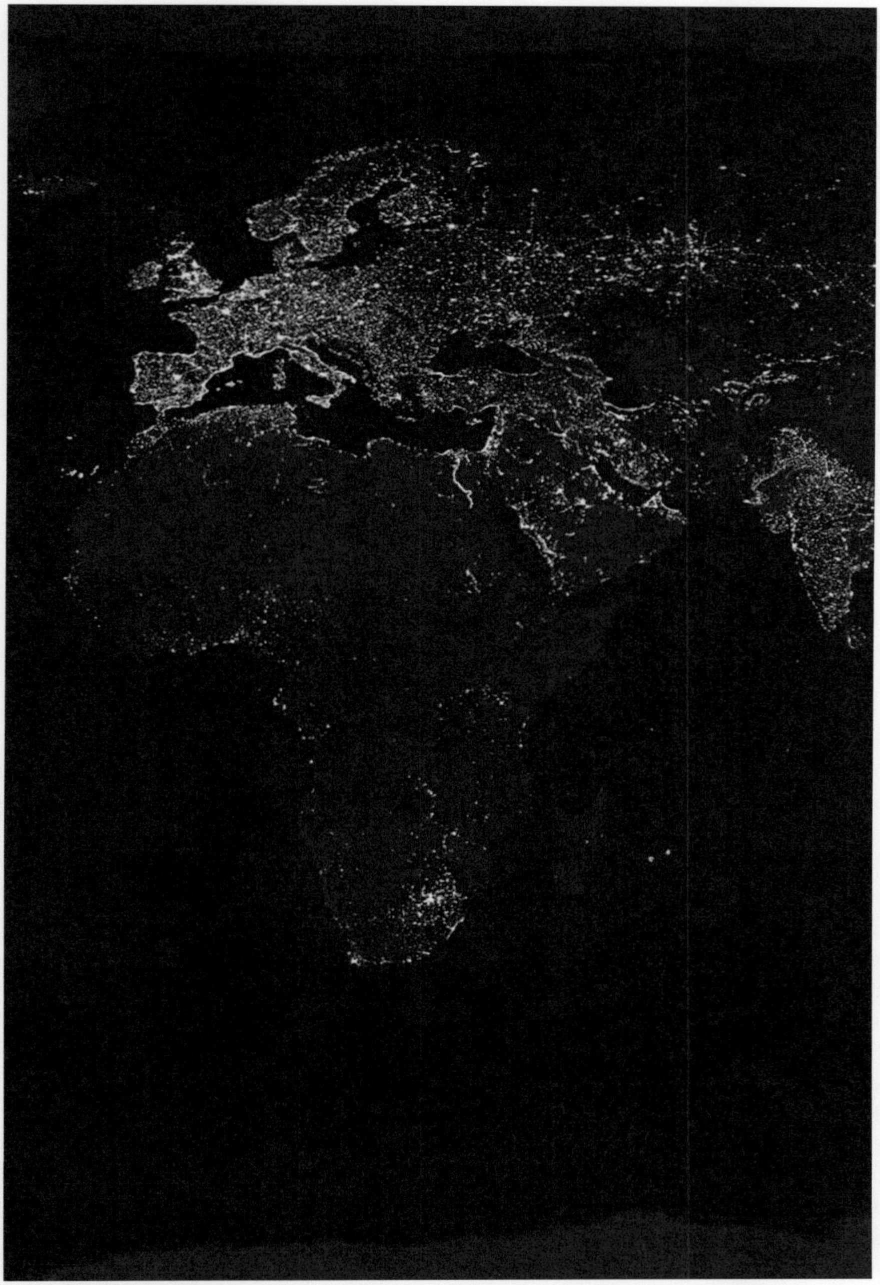

(continued)

◀ *Concentration to resources.* This photograph, taken from space, shows Earth's night-time illumination. The River Nile, with its fertile basin, appears as a root in North Africa. Cities dot the map as bright spots. The black areas—oceans and deserts—are more or less devoid of human activity. The photo could also represent biological activity in the soil. Life in the ground, like human life, is concentrated in areas where resources are plentiful. A fallen leaf provides easily accessible nutrition for fast-growing bacteria and fungi. Plant roots leach sugar, and here the activity level is hundreds of times higher than a few millimetres away from the root surface. The *dark areas* could represent mineral grains that are resistant to weathering or areas with old organic matter depleted of accessible nutrients

can survive on a resource composed of carbon, oxygen, hydrogen and little else. Predators eat prey rich in protein and with a similar chemical makeup to themselves, but how do termites ingest enough nitrogen, and where does the surplus carbon go when they convert their food into biomass? The answer is that termites, like all other herbivores, exist in symbiosis with microorganisms in their gut.

I learnt how to study these microorganisms at a conference in Halifax, Canada. You can dissect a termite by removing the head with tweezers and removing the alimentary canal with a needle. You then squeeze out the contents of the gut in a drop of water, put them under a cover glass and examine them with a microscope. You will then observe a myriad of small organisms of all shapes and sizes swimming around in the water—long, corkscrew-shaped creatures wriggling alongside oval-shaped microorganisms that use small rotating hairs for forward motion.

The former are spirochetes—giants of the bacterial world that can grow to a tenth of a millimetre in length (almost one hundred times longer than most other bacteria). They ferment carbohydrates decomposed from wood fibres in the anaerobic conditions prevailing in a termite's stomach. Some spirochetes are known for causing illness in humans, including the tick-borne Lyme Disease.

The oval organisms are unicellular ciliates. They also feed on carbohydrates and contain archaea that convert the carbon dioxide and hydrogen produced during fermentation into methane. Thus, excess carbon, oxygen and hydrogen in termites' plant food is released into the atmosphere as carbon dioxide and methane. By eating large quantities of wood, termites ingest sufficient nitrogen for their needs.

Bacteria and ciliates gain substantially from this symbiosis. Theirs is a sheltered world offering protection from predators and unfettered access to food. For their part, the termites do not have to compete with other animals for food as they alone are capable of eating wood.

Termites are widespread in tropical and subtropical areas. Anyone who has been on safari in Tanzania's Serengeti National Park will recall the ubiquitous sight of tall termite mounds. The insects build labyrinths of underground tunnels and drag in bits of wood and other organic matter when collecting food for the colony. The mound is literally just the tip of the iceberg. In some savannah areas, termites dispose of a quarter of all fallen leaves, and play the role of composters just

Soil invaders. The earthworm is a kind of ecosystem engineer: its activity has major structural consequences. Nutrition circulates faster and primary production increases when worms remix the material in the soil. But if the ecosystem is not suitable for this activity, the consequences can be devastating. Field vegetation disappears almost entirely in the deciduous forests of the northern United States (*lower image*) when European earthworms invade the soil. The upper image shows a normal forest without worms

Perfect temperature. Termites are masters at building cathedral-like mounds with advanced ventilation systems—a necessity in the hot African savannah and Australian bush. Some termites cultivate fungi on organic material that they bring home and store in underground passages. Fungal cultivation fails in excessive heat so warm air is vented through chimney-like channels that open out at the top of the mound

as earthworms do elsewhere. The compost enriches the soil and patches of lush vegetation can often be seen growing in and around abandoned termite mounds.

Termites and Their Cultural Legacy

In Australia, an Aboriginal musical instrument, the didgeridoo, is a eucalyptus log hollowed by termites. When blown, the didgeridoo emits a deep vibrating tone that varies in timbre according to the position of the mouth. It is also ideal for imitating animal calls. My interest in the didgeridoo was aroused on a visit to Kakadu National Park in Australia's Northern Territory. Eucalyptus trees stood in sparse groves wherever you looked, and if you tapped on the trees they sounded hollow (many would doubtless have made perfect didgeridoos). Termite mounds the height of a fully grown human stood scattered between the trees—creations of densely packed red earth mixed with undigested residues from termite faeces. Nothing grew

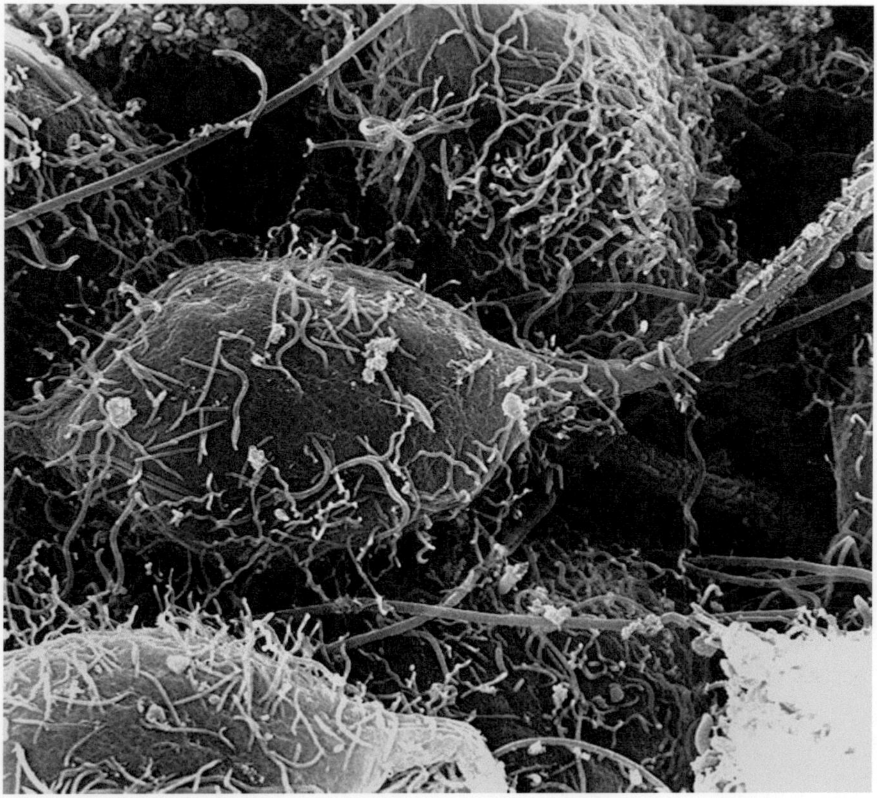

Security of abundance. Termites' digestive systems are full of spiral-shaped bacteria (spirochetes) and single-celled ciliates. These microbes break down wood and the termites survive on the residues. In exchange, the microbes get abundant food and protection from predators. The photo shows microscopic spirochetes wrapped around the gigantic ciliates that can measure up to a tenth of a millimetre in length

beneath the trees due to a toxin in eucalyptus leaves that prevents seeds from germinating and plant-inhibiting substances secreted by the trees' roots. The toxin is a muscle relaxant and lowers blood pressure and body temperature, which is why koalas—which exist strictly on a diet of eucalyptus leaves—are so slow and ponderous when moving through the canopy. In return, the koala avoids parasites such as lice and ticks as these eschew the toxins secreted in the animal's skin.

Aboriginal music is heavily influenced by the natural world. Aboriginal players use their didgeridoos to imitate the calls of birds and other animals and also impersonate animals' behaviour in their dances. I have enjoyed wonderful moments playing my didgeridoo to fire-bellied toads and edible frogs at different locations in southern Sweden. The low-frequency vibrations from the instrument excite the

Others do the job. Termites have access to a unique food source—wood—through their cooperation with microbes. Australia's aboriginal people exploit this phenomenon when making their traditional musical instrument, the didgeridoo. A didgeridoo is made from a eucalyptus branch eaten hollow by termites

amphibians and the males answer immediately. Female frogs and toads favour males with a deep croak as this is a sign of physical health. On one occasion the females came so close I thought they would start crawling up my leg.

Further Readings

Baskin Y (2005) Under ground. How creatures of mud and dirt shape our world. Island Press, Shearwater Books, Washington DC
Darwin C (1881) The formation of vegetable mould through the action of worms with observations on their habits. John Murray, London
Dexter Dyer B (2003) A field guide to bacteria. Cornell University Press, New York
Marshall J (2007) War of the worms. New Sci 3
Smith S, Read D (2008) Mycorrhizal symbiosis. 3rd Edition. Elsevier Academic Press, San Diego
Young S (2011) Craving earth. Columbia University Press

The Nutrient Cycle

Compost—that sensual kingdom where
fetid waste is refined into fragrant
humus, to serve our growing friends.

Lars Krantz

Abstract The circulation of nutrients is essential to the survival of all living things. The supply of water and oxygen, the temperature and acidity in the soil are of great significance to this process. In a tropical rainforest, the cycle is rapid because the temperature and humidity are just right. Nutrients are accumulated in the vegetation rather than in the soil. In the boreal forest, the reverse is true. Low temperatures and low pH mean that nutrients are accumulated in thick organic layers, which cause the trees to grow slowly. Experiences from natural systems can be useful when setting up a compost heap in your own garden. Along with fact-based texts, the chapter contains my own observations of the fascinating work of dung beetles and coprophilous fungi. I conclude with a reflection on sustainable gardening and how it may have been inspired by the conditions in the rainforest.

It is not easy to understand why organic matter collects in some soils and vanishes quickly from others. When researchers asked primary school children what happens to fallen leaves in autumn, they found that most of the pupils understand that leaves break down in some way or other and become part of the soil. Some of the children reckoned that the addition of new soil every autumn ought logically to lead to the Earth growing larger. They had yet to understand that during decomposition some matter enters a different phase and is released into the atmosphere as carbon dioxide, while certain minerals dissolve into liquid form in the soil and are absorbed by plants. Likewise, many of us may not stop to think that waste incinerated in a power station does not simply become a pile of ash, but that most of it re-enters the atmosphere as carbon dioxide and other gases. In one respect, though, the schoolchildren were right to suggest that the amount of soil in the world ought to rise. For in some places, soil does indeed accumulate, albeit slowly, if left undisturbed—and may over time grow into deep layers.

H. Wallander, *Soil*, DOI 10.1007/978-3-319-08458-9_5

My Compost Bin

Organic matter can decompose rapidly in a compost bin, especially when fresh, and can lose up to two thirds of its weight in the first few weeks when fungal and bacterial processes are at their most active. But getting the right balance in your compost bin is not always easy. If the mix is too damp it can smell sour, and you will find the compost sticky and poorly decomposed when you rake it out from the bottom. After tipping food scraps into my compost bin I line the bottom of the bucket with a thick layer of sawdust to absorb excess moisture before putting it back in the kitchen.

The modern compost bins sold for household waste are actually not ideal for the job. Near-hermetically sealed, they create an ambient environment with too much moisture and too little oxygen. One way to overcome this is to place your bin on a pallet and allow excess water to run out, provided there is a drainage hole in the bottom. I collect this runoff, mix it with tap water and use it to water my tomatoes. Ideally, you should stir or turn the compost every now and then to oxygenate it. Some compost bins are designed to rotate, which makes this easier.

When composting nitrogen-poor materials like straw and twigs, it can be difficult to get the decomposition process going. Microorganisms need nitrogen to reproduce, and poor aeration may halt the degradation process. Conversely, kitchen waste containing a lot of food scraps can contain too much nitrogen. Here, sawdust—which contains much more carbon than nitrogen—plays a balancing role. Any excess nitrogen may run off as nitrate or seep out as ammonia or nitrous oxide, though this is detrimental because nitrous oxide is a greenhouse gas and ammonia and nitrates cause eutrophication. Leaching of these substances also deprives the compost of important fertiliser and defeats the object of the exercise. Better, then, to bury the waste in the vegetable patch so that growing plants can absorb the excess nitrogen.

A pumpkin grower called Börje Gustavsson has developed a special drill for precisely this purpose. Pumpkins are a demanding vegetable that requires aerated soil, heat and plenty of nutrients and water to grow to a good size. Food scraps buried near the plants provide nutrients, aerate the soil and create heat when decomposing. Börje, who won the Swedish Pumpkin Championship in 2003 with a monster weighing four hundred and fifty-three kilograms, reveals on his website a key ingredient in his success: a urine-diverting toilet. I have tried using the drill in my garden but as yet not managed to grow a pumpkin as large as Börje's. It looks like the next step will have to be using urine as fertiliser if I am to reach the upper echelons of the pumpkin-growing world.

My best composting efforts came as a student at Capellagården School on the Baltic island of Öland, where I took a gardening course and learned to take the right amount of chicken fertiliser (for adequate nitrogen) and mix it with twigs from old raspberry bushes (to provide aeration), weeds that had not gone to seed and food scraps from the school kitchen. I covered the compost heap with a thick layer of straw and stuck a plastic tube into the middle to provide aeration. Heat and a good oxygen supply are vital for effective decomposition. I used to enjoy watching the steam rising from the plastic tube on my morning visits to the chicken coop to collect the eggs for breakfast.

Recyclers seldom sleep. One might think that all decomposition stops during winter, but fungi and bacteria can continue their activity under the snow when the temperature remains around zero. Decomposition is slow when it is cold but seldom stops altogether and there is no risk that conditions will be too dry under the snow. But if water saturates the ground, expelling oxygen, decomposition proceeds much more slowly. This is why marshes can have peat layers several metres thick

Dry goods. Decomposition cannot occur if conditions become so dry that bacteria and fungi lack water—a fact humans have exploited since ancient times to preserve food. People in this Nepalese village hang out corn cobs to dry in the sun so they can later grind the kernels into flour. However, it is often difficult to keep foodstuffs dry enough to prevent mould and bacteria growth, and a large part of the global harvest is destroyed because of inadequate storage

Achieving a well-balanced compost heap is about creating optimal conditions for bacteria to grow. When that happens, the temperature inside the heap can exceed seventy degrees Celsius. That level of heat kills most parasites and weed seeds, saving you all sorts of problems when you put the compost on your growing beds. The temperature usually peaks at around one week and then gradually falls, at which point fungi start to play their role in the decomposition process. Once the temperature drops to between 35 and 40 °C, different species of ink cap mushroom get to work breaking down hard-to-digest lignin. They also attack other fungi to ensure exclusive access to the compost material. Field mushrooms will grow at temperatures below 30° but you seldom see them on compost heaps because ink caps attack their mycelia and prevent them from sporulating. To grow field mushrooms one must therefore pasteurise the organic matter and inoculate the mycelia.

At the Tiputini River in the Amazon

Compost heaps are not the only place where fungi thrive. They are also very much at home in tropical rainforests. I travelled to the Amazon basin in eastern Ecuador with a group of scientists specialising in symbiotic organisms. Some worked on lichens, a symbiosis between algae and fungi, others on the symbiosis between termites and their intestinal bacteria, which enable them to eat wood. For my part, I went to study mycorrhiza, a symbiotic relationship between fungi and plant roots.

Active clay. The Tiputini River leads deep into the Ecuadorian rainforest. Here, the water level can fluctuate by 11 m. When the riverbanks are exposed, parrots gather to lick the clay at favourite spots. The clay functions as active carbon, binding to and neutralising toxins in the seeds that the parrots eat

Our guide was Meyer, a descendant of Spanish colonists and an employee at the research station on the north bank of the Tiputini River in the Ecuadorian interior close to the border with Peru. On the southern bank is Yasuni National Park, one of the areas of richest biodiversity in the world, with more than one-and-a-half thousand species of tree, five hundred and twenty species of bird and countless insect species. Yasuni National Park is also home to the Huaorani people, an isolated tribe of indigenous people who speak a language unrelated to any other known tongue. Meyer was well acquainted with their customs and traditions and had learnt many secrets of rainforest life from them. Little escaped his eagle eyes during our treks in the jungle.

On our first hike, Meyer tried to get us to creep quietly forward, listening to the jungle soundtrack and picking up the scent of a jungle pig that had just crossed our path. Most animals fled as soon as they detected our scent, but occasionally Meyer managed to pick up a scent before the animal caught ours. The monkeys were less shy as they swung around in the treetops, and brilliantly coloured parrots shrieked hoarsely as they flew overhead. The dense rainforest enveloped us in a warm and humid embrace, though every now and then the undergrowth would thin out and the ground might even be laid bare. I fantasised about these clearings being places where rainforest dwellers held councils and where their footsteps had kept the vegetation at bay. On one occasion we stopped at a glade with just a few straggly trees and Meyer drew our attention to the swollen tips of the branches. When he cut one with his knife, an army of ants came swarming out. He asked if anyone wanted to have a taste, and when no one volunteered he stuck his tongue into the branch until it turned black with ants. Meyer sucked up the acidic juice with a mischievous smile.

Scraped bare. Look up into the rainforest canopy and you will see many insect-ravaged leaves. Plants have developed various strategies to deal with this problem. Some produce toxins to reduce insect attacks; others focus their resources on producing many leaves to offset foliage loss

Thirsty sentinels. Some plants cooperate with ants to protect their leaves. The ants attack all potential intruders and are repaid in the form of a high-energy fluid from special glands in the leaves

"Lemon ants," he said. "They're delicious. The Indians call them the devil's gardeners. Their acid prevents weeds from growing round the tree trunks and the trees thank them by letting them live in their branches."

After a few seconds' hesitation, we cupped our hands. Suddenly, everyone wanted to try the lemon taste and hear Meyer tell us about the indigenous people's millennia of knowledge and the rainforest's plants and secrets. And when he picked up a fallen fruit and showed how its juice could be used to colour the skin, we all wanted our hands and arms decorated with Huaorani patterns.

In Rainforests, Decomposition Takes Place So Fast that New Soil Has Virtually No Time to Form

Rays of sunlight pierced the treetops. The leaves had more holes than a Swiss cheese thanks to the voracious appetite of insect larvae, whose excrement in turn falls to the ground and is quickly broken down by fungi and bacteria. Leaves that survive intact eventually fall to the ground too, along with twigs, blossoms and flowers. It is never totally quiet in the jungle: you can always hear the sound of something falling. Some leaves do not make it all the way to the ground but lodge in crevices and decompose into soil high above. Plants whose seeds germinate in these hollows spend their entire life marooned from the ground below.

◀ *Recycling workers*. Decomposer saprotrophic fungi are quickly in place and grow their mycelia to break down withered leaves. Here (*top photo*) mycelia have grown out from a thin twig. Fungal spores on the surface are ready to grow when the leaf dies. Some fungi grow in living leaves without damaging the plant, ensuring they are first on the scene when the leaf dies. Many different kinds of fungi together break down leaves in the rainforest. The middle photo shows a pyrenomycete, a relative of the stag's horn fungus that grows on dead tree-stumps in Europe. Its antennae-like fruit bodies facilitate spore dispersal. The thin leaf (*lower image*) contains sufficient energy for the fungus to build a forest of pin-like fruit bodies. The spores spread to new leaves and the fungus can live on while the cycle of growth and decomposition continues

Everywhere I looked, empty pods or chewed fruit dotted the forest floor, signalling the presence of monkeys. A thin layer of brown leaves covered the ground. Hold the leaves to your face and a full-bodied mushroomy smell hits your nostrils. The leaves do not have time to build up on the forest floor because the heat and humidity create a rapid decomposition process of just a few weeks. Indeed, we did not see any ground layer of organic matter anywhere in the jungle. Beneath the leaves lay a tight blanket of delicate tree roots colonised by mycorrhizal fungi that absorb the nutrients of decomposing leaves. The soil beneath lacks sufficient nutrients to support the plants and trees by itself and relies on the continuous addition of leaf litter. When rainforest is clear-felled the nutrients quickly disappear and the soil can only support a few harvests before becoming too depleted for farming. It takes a long time for such soil to become fertile again.

An unknown beauty. The rainforest contains many species not yet known to science. The unusual structure on the right of the photo remains unexplained, though I have consulted several experts in tropical mycology. Probably it is a kind of vegetative propagation organ that lies on the ground waiting for organic matter to fall from the trees

Towards new goals. Wood breaks down fast in rainforests, where constant humidity and heat provide perfect conditions for decomposer fungi. In northern forests, fallen tree trunks can remain on the ground for centuries, whereas in the rainforest they disappear in a few years. The image shows fungal fruit bodies sticking up from a rainforest tree trunk like knobbly black fingers. A similar species in Europe and North America is called dead man's fingers

During the second day's hike I spotted a palm leaf hanging like a dirty rag over a liana. The greyish-white leaf was covered in fungal mycelia that caused it to disintegrate into powdery scales when I touched it. On the edge of the leaf grew small white fruiting bodies that had already obtained the nutrition they needed to emit spores that would disperse to find new organic matter to colonise. The indigenous people uses these palm leaves to weave baskets in which they carry their few worldly possessions on their rainforest wanderings.

Mushrooms are everywhere in the rainforest. Lovely clusters of small brown fruiting bodies with sparse hyphae connected by brown threads carpeted the enormous plank root of a kapok tree, the rainforest's majestic giant. The mushrooms were of the genus *Marasmius* and turned my thoughts to Sweden's spruce forests, where carpets of needles can be covered with tiny stinking parachutes, a close relation from the same genus. Pick up a stinking parachute and it will be attached to a spruce needle that it is breaking down. In Sweden, though, the decomposition process is slow; even after ten years, a third of the needle's mass may remain as a permanent residue.

Giant Trees Topple in the Dead of Night

A huge crash woke me in the middle of the third night. A kapok tree had fallen, bringing down several other trees nearby. Rainforest biodiversity benefits from the disruption caused by falling trees, and the resultant clearings create space for more

Future fungus food. The Tiputini River is hemmed by a green curtain of rainforest on either side of the river. Monkeys and colourful birds like to visit the spreading crown of the kapok tree, the tallest in the forest

light-dependent species. I later saw the remains of a kapok tree that had fallen long before: the glade in the forest had closed up and all that remained of the tree was a mossy stump of plank roots. I could see where the trunk must have lain but could find not a single remnant of wood: everything had rotted away. Here and there I saw depressions in the ground, like a giant's footprints, that Meyer told me were left by trees that had decomposed right down to their roots. The contrast with the forests of Sweden, where it takes many years for tree stumps to revert to moss-covered mounds, could not have been starker. Tree stumps in the northern part of the country can take hundreds of years to decompose due to the powerful preservative properties of pine resin.

Another example of ultra-slow decomposition is the Bocksten Man, the remains of a medieval man buried in a peat bog in Varberg, western Sweden. The man had been murdered, with stakes driven through his body by his attackers in the belief they would prevent his spirit from haunting them. The stakes still remained well preserved after more than eight hundred years in the bog. Compare that to the rainforest trees that vanish without trace—bar a hole in the ground—in a few short years. The contrast is explained by oxygen availability, which regulates the pace of the decomposition process. Oxygen struggles to penetrate damp bogland, just as it does a wet compost heap. Anaerobic bacteria work slowly and fungi struggle to survive in oxygen-deprived conditions. Bog environments are also acidic, further slowing the decomposition process.

Spear-Shaped Spores and the Huaorani

The day after the night-time crash, Meyer discovered an orange club-shaped fruiting body growing on a rotting tree trunk. He carefully cut into the wood with a knife to reveal a large insect covered with mycelia and with the fruiting body protruding from its abdomen. It was under attack from a *Cordyceps* fungus specialised in breaking down dead insects. The fungus, which mobilises an arsenal of enzymes to break down the hard shell, is believed to have potent pharmacological properties and is used in traditional Chinese medicine. Western scientists have also shown interest in the substances secreted by this and other *Cordyceps* species.

Back at the research station we put the dead insect under the microscope, where, once it had warmed up under the heat of the microscope light, we observed javelin-shaped spores being emitted from the fruiting bodies. Sickly insects often succumb to the spores and may carry a fungal infection for a long period before eventually dying. It is said that infected insects isolate themselves from their congeners and may cover long distances to avoid spreading infection.

The spores made me think of the Huaorani people in the rainforest near the station. They have never been conquered—by the Incas, by Spanish conquistadors or by other larger indigenous tribes. They are famous for their sharp spears, made from the hardest wood and reputedly capable of killing fifteen men without

Potent spears. Cordyceps fungi growing out of rotten tree trunks reveal where insects lie buried. The insects can carry the fungal infection for a long time but eventually die, whereupon the fungi build stately clubs to spread their spores to new insects. The photo shows our rainforest guide carefully cutting out the dead insect with his knife

breaking. For centuries, the Huaorani were feared warriors, as the first foreign missionaries discovered to their cost. Today, the tribe's worst enemy is the international oil industry, which is advancing farther and farther into its territory. Indeed, we could hear the distant drone of oil exploration generators from the research station, deep in the rainforest.

On my mornings beside the Tiputini River, as I looked across to Yasuni National Park on the opposite bank and listened to the strange sound of howler monkeys through the mist, I wondered how far inside the lush curtain of trees the Huaorani were. They had little contact with the research station staff, despite living so close, though when the centre was built they came with a tame grey-winged trumpeter as a gift. The bird was named Lucy and turned up each lunchtime for a bowl of rice and fruit. We took it in turns to stroke her head after she had finished and her presence—together with the Huaorani spear hanging on the canteen wall—stiffened our resolve to fight the oil companies and protect the forest. The more scientists who come here and raise awareness about the area within the research community, the greater the chances of stopping the advance of big oil.

Moi, a Huaorani spokesman, wrote to the US president inviting him to visit the tribe in its rainforest home. He wanted the head of the world's largest economy to explain why oil companies were destroying the Huaorani's land and why missionaries wanted to civilise them. The Huaorani simply want to continue living as they always have: in harmony with the spirit of the jaguar. Moi travelled to Washington, DC—a journey that took him two weeks—to deliver the letter in person, but the White House gates remained firmly closed to him.

The jaguar is the most mythical of the Amazon's inhabitants. Few have seen one in the wild, but there is no doubt that a healthy population exists in and around the research station compound, where they walk the same trails as we did. Camera traps identified nine jaguars in the station's immediate vicinity. Perhaps it was just as well that we never encountered one.

Dung Beetles and Mushrooms: Important Players on the Decomposition Stage

We found the prints of various animals on the muddy tracks near the station but never any droppings, despite thorough searches. We wanted to observe dung beetles, the custodians of rainforest hygiene, rolling faeces and burying them in their burrows. But droppings are a precious resource in the rainforest and there is stiff competition to be first on the scene.

One evening about a week into our stay I was sitting in the research station library reading *Tropical Nature* by Adrian Forsyth and Ken Miyata, from which I learnt that human faeces are the most popular dung among rainforest insects. My dung beetle studies received new impetus and I decided to conduct an experiment: I would leave my own deposit in the forest—not too close to the trail but not too far into the forest and its battalion of poisonous snakes and spiders—and see what

◀ *Luminescent fungi*. Aristotle observed that certain fungi emit a mysterious green light at night. But even today, we are not sure why. Perhaps the fluorescence comes from the by-products of wood-decomposing fungi. It could also be a way to attract nocturnal flying insects as spore dispersers. Another possibility is that the light attracts predators that eat mosquito larvae and other bugs in the fruit bodies. The light emitted is weak, but can be caught on camera at slow shutter speeds

happened. Perhaps the greatest risk was of irritating a bullet ant when squatting down. Meyer had been in pain for days after being bitten by one in the arm, and the symptoms would doubtless be even worse from a bite to more delicate parts.

As it happened, the dung-laying process went without a hitch and I was able to view the action with binoculars from a comfortable distance. Back at the station I continued my reading and discovered that I could expect rapid colonisation by egg-laying flies, but that it could take a few hours for the first dung beetles to arrive. If I was lucky, my excrement would attract rhinoceros beetles that would fight violent battles with their horns for the right to lay their eggs there. Rhinoceros beetle larvae live on maggots that hatch in the excrement. Clearly, I had quite a spectacle awaiting me, though I decided to keep the experiment to myself initially and not tell any of the other scientists. I returned after an hour and noted large blowflies dining on the dung. One had a long neon-coloured abdomen that shone like a gemstone against the dark forest floor. But there were no dung beetles in evidence yet.

Profitable enterprise. Elk or deer droppings are a familiar sight in Scandinavian forests. In rainforests, however, animal excrement is rarely seen. Here, dung is an important resource and the males of around fifty different kinds of dung beetle compete to roll balls of droppings. When the balls are ready, the beetles perform a characteristic dance to register their surroundings. Then they roll the balls away and hide them from other males. If the beetles encounter obstacles along the way, they perform the dance again to orient themselves and make sure they are heading away from the dung pile

My second visit, another hour later, revealed nothing new and when, after four hours' observation, the only progress was the presence of a few large cockroaches, I started to feel relieved I had not told my companions of my impromptu scientific probe.

Darkness falls quickly in the rainforest and you scarcely have time to register that the light is fading before pitch blackness descends. I decided to make a final visit to the dung heap and entered the forest with a torch. As I walked I heard male tree frogs calling to attract a mate. The frogs sit high up in the canopy, where females lay their eggs in the water-filled rosettes of bromeliads. When the tadpoles become frogs they crawl out into the foliage in search of pools of their own.

At my feet, tiny fluorescent spots shone on the ground when I turned off my torch, reminding me of old-fashioned light switches that glow so you can find them in the dark. Carefully trying to avoid touching a bullet ant or, even worse, a poisonous spider, I rummaged around trying to find out what they were. Earlier that day, Meyer had pointed out a spider with a bite that can kill a man in three minutes. Eventually my fingers located something, and when I turned on the torch I saw it was a leaf—a leaf illuminated by florescent fungal mycelia. Why this particular fungus has developed an ability to shine in the dark is not known.

On reaching my vantage point I shone the torch on the dung heap and saw that there was activity at last. Dung beetles were working hard to turn my faeces into balls that they then rolled by kicking with their hind legs, like squirrels on a treadmill, while keeping their other legs on the ground. Balls of dung were being rolled left and right, and after a few metres negotiating leaves and twigs the beetles would start burying them in readiness for female dung beetles to enter the scene. Males who lay the most and largest balls have the greatest chance of fertilising the eggs that the female lays inside them. After fertilisation, females remain behind to tend the balls, turning them regularly to prevent mould from forming, while the males amble off in search of new dung deposits. More than fifty species of dung beetle have been found in human dung in jungle environments according to *Tropical Nature*, and I had seen just a fraction of these on my pile. When, the following morning, I took my colleagues to see the experiment, there was no sign of either dung or beetles. But I contented myself with the knowledge that my presence had given the local insects a helping hand—for a few days at least.

The Swedish biologist and author Fredrik Sjöberg has a special interest in dung beetles and in his anthology of essays *Den utbrände kronofogden som fann lyckan* ("The Burned Out Debt Collector Who Found Happiness"), he describes a case from Australia that highlights the vital role insects play in the ecosystem. When European colonists settled in Australia, they brought with them cattle from home for beef farming. However, the indigenous dung beetles ignored the alien dung, which was left to collect in the fields and became a magnet for the larvae of biting insects. The insects reached plague proportions, rendering vast areas of land useless for livestock farming. Farmers responded in the mid-twentieth century with a novel experiment: importing dung beetles from Europe and Africa. The gambit proved successful and the cycle of dung production and decomposition now functions well in the Australian outback.

Dung plays host to all sorts of interesting biological phenomena. When teaching a distance course in microbial ecology, I tasked my students with studying

Explosive stems. The *Pilobolus* fungus genus often grows on herbivore droppings. It has spore-containing vesicles that turn towards the light and eject spores at a speed of 25 m/s—an acceleration rate faster than a bullet from a rifle. If the spores land on a blade of grass, they can eventually end up in a horse's stomach. After passing through the animal's digestive tract, they grow directly in the fresh manure

the fungi that appeared on a fresh pile of horse manure. I directed them to put the manure in a sealed transparent container lined at the bottom with damp tissues. The students made daily observations, described all the fungi and took documentary photographs. After a few weeks I received a detailed report complete with photos of sporulating *Pilobolus* covering the entire manure pile. *Pilobolus* is a fascinating fungus that subsists on the most readily available nutrients. Its strategy is to be a first mover and to grow fast. The horse ingests *Pilobolus* spores attached to blades of grass, and the spores germinate directly after passing through the horse's gut. *Pilobolus* then thrives for a brief period after the horse excretes the manure before giving way to other fungi, which are left to exist on cellulose fibres and hard-to-digest lignin. Meanwhile, the *Pilobolus* orients itself to the light before emitting its spores in pressurised blasts that are the fastest recorded motion in the natural world, reaching speeds of up to 25 m/s. Ideally, the tiny spores land on vegetation that will be grazed by a new host. *Pilobolus's* ability to orient towards the sun has inspired an American dance company to invent a new type of modern dance. Check out Pilobolus on YouTube and enjoy fungus-inspired choreography.

Scandinavian Cold Enables Humus Layers to Grow Thick

Leaves and needles that have not decomposed fully can accumulate to form thick layers of soil. This happens if the temperature is sufficiently low to prevent full decomposition of fallen leaves and twigs before the process starts again the following autumn. I saw this with my own eyes on a trip to northern Sweden.

The summit of Akkeli mountain lies midway between lakes Uddjaure and Hornavan near the town of Arjeplog in Swedish Lapland. The sweeping view from the peak takes in a myriad of islands, most of them covered in forest and largely untouched by human hand. The islands' inaccessibility makes exploitation for logging unprofitable, in stark contrast to the clear-fells and young plantations that pockmark the mainland. Not all the islands are untouched, however: some have red wooden cabins with green lawns. Here and there are patches of bright purple, where fireweed has taken hold. This stately plant is common throughout most of Sweden, especially in the north.

Along with Professor David Wardle and his colleagues at the Swedish University of Agricultural Sciences, I visited some of the islands to take samples. Together we filled a motorboat with soil augers, insect nets, collection jars and quadrats. Scientists have taken an interest in the islands ever since a forester in the 1930s noticed that several of the smaller ones had much thicker soil than typically found on the mainland, with a layer sometimes more than a metre thick on the scree that constitutes the geological substratum. So thick was the soil that by the water's edge it formed banks high enough for sand martins to nest in. The forester reported his findings to the National Council for Forest Preservation, but they did not come to the attention of university researchers until many years later. David Wardle and colleagues from all over the world now visit the islands every summer to take samples and record data.

Our first port of call was a medium-sized island dominated by birch trees, with lesser numbers of pine and spruce. Lingonberry, bilberry and crowberry bushes covered the ground, though the berries had yet to ripen. I settled down on the shore and watched the augers drill into the soil and the insect nets sweep through the undergrowth. Within minutes I was covered in ants and, as I jumped up frantically trying to dust them off, I gave the insect net-wielding Michael Olsson a good laugh. He called the place "Ant Island" and told me he had caught more than twenty thousand ants in his traps there, compared to just a few hundred on the other islands.

"Why is that?" I asked him.

Michael had no firm answer, having been unable to establish any link between the ants and the type of vegetation or island size. He had, however, noticed that aphids were commoner and spiders less common compared to the other islands. Ants are known for "milking" aphids for the honeydew they secrete when eating plant sap, and everywhere I looked I saw ants scaling the birch trunks to reach aphids in the leaves and others descending, laden with honeydew to take back to their nests. The ants show their appreciation for the aphids by protecting them from spiders, which they attack and kill.

We moved on to the next island, which was much larger and dominated by pine and bilberry bushes, on which the fruit was ripe. I gorged on the berries while the

Waiting for fire. The likelihood of lightning striking and causing a forest fire is much lower on small islands than on large ones. Applying this probability, researchers have studied how fire frequency affects soil processes on different-sized islands in the Uddjaure and Hornavan lakes in northern Sweden. The humus layer is only a few centimetres thick on the large islands, as in standard forest soil, but can reach a depth of 1 m on the small islands

scientists took their samples. The forest was in fine condition and it occurred to me that unscrupulous-minded people might be tempted to fell a few trees and float them over to the mainland. Fortunately, no such incidents have occurred and the islands remain undisturbed and a scientists' paradise. There is a clear link between tree growth and the islands' size, with trees growing best on the largest islands and vice versa. Spruce dominates the canopy on the smallest islands and crowberry the undergrowth. Fallen spruce needles and crowberry leaves decompose slowly and the soil here is so deep because the addition of organic matter outpaces the rate of decomposition—a process that has continued at an annual growth rate of two millimetres since the Ice Age and which will continue so long as the islands remain undisturbed. On the smallest islands, the soil reaches a depth of more than one and a half metres.

The islands have never been immune to disruption—even before humans set foot there. Fierce forest fires would burn away the topsoil and signs of old blazes can be seen on some of the ancient pines, especially on the larger islands. This is because the larger the island, the higher the chance of a lightning strike. The bigger islands experience blazes on average once every five hundred years, and because pine is equipped to withstand forest fires (its thick bark offers decent protection against flames) it predominates there. Pine seeds that germinate after a fire grow especially well because the nutrients from the humus are concentrated in the ash. Bilberry bushes also benefit because they grow faster than lingonberry and crowberry in nutrient-rich soil. Fire frequency is much lower, at around once every three thousand years, on the smaller islands and here spruce predominates, while lingonberry and crowberry are commoner in the under-storey because they fare

Seizing opportunity. Rosebay willowherb, also known as fireweed, is originally a forest-dwelling plant whose easily disseminated downy seeds can survive for long periods in the ground. The seeds germinate quickly after the ground is disturbed. Before human intervention began to change the forest landscape, it was mostly found on burnt ground, as here in Muddus national park, where blackened pine trunks bear witness to the major fire that swept the park in 2006. Now the plant is common in clearings and along roadsides. It got its other colloquial Swedish name—railwayman's rose—from the fact that it followed in the wake of the railway workers as the tracks were built

better than bilberry in poor-quality soil. This, in turn, promotes humus accumulation in the topsoil as spruce needles and crowberry leaves decompose more slowly than their pine and bilberry equivalents.

Vegetation thrives in the aftermath of a fire. Plants are quick to exploit the nutrient-rich ash, though the charcoal left behind by a blaze can remain intact for thousands of years. Charcoal can be aged using carbon 14 dating. By examining traces of fires on old trees and sifting the charcoal from soil samples, scientists have charted the islands' forest-fire history. The carbon remaining after a blaze has an impact on numerous soil processes. Carbon can bind substances that inhibit microbial activity, thereby stimulating nutrient cycles. Nutrients attach to the charcoal's large active surface area and thus avoid being washed away by rain. This is why charcoal is used to neutralise contaminants ingested during food poisoning. A bag of activated carbon is dissolved in water and given to the patient to drink. The toxins attach to the carbon particles and can then pass harmlessly through the alimentary canal.

I saw extensive evidence of fire on the ancient island pines, though it was not until I visited Muddus National Park—300 km to the north—that I witnessed the ravages of a fire at first hand. The southern part of the park is stony and the primary pine forest has experienced regular fires every two to three hundred years. The forest is magical, with large, moss-clad boulders, twisted tree trunks, lichen-draped branches and a soft carpet of bilberry and lingonberry bushes. Then, all of a sudden, a disturbing clearing of bare boulders and blackened trunks appeared: the site of a fire two years earlier. At first glance the ground looked dead, but on closer inspection there were signs of life. Tiny bilberry shoots were poking through the ash and green lichens were colonising the damper areas. Most impressive of all was the fireweed, its vivid blooms a splash of colour among the blackened trees. As its name implies, fireweed is quick to colonise burnt ground after forest fires, taking advantage of the absence of competing vegetation.

When the Grass Really Is Greener?

Everything ultimately decomposes and degrades into gases and minerals. Sometimes conditions are not optimal and the process takes longer, as in the case of the Bocksten Man. By contrast, the decomposition cycle is fastest in rainforests, where plant matter breaks down so fast that humus has no time to form. The combination of sparse soil and rich vegetation found in tropical forests reminds me of Nils Åkerstedt, the father of cover-crop growing. In his books, Åkerstedt claims it is possible to grow anything in nutrient-poor soil provided one applies a thick top layer of grass clippings. With their perfect blend of carbon and nitrogen and good aeration, grass clippings facilitate rapid decomposition—mimicking rainforest conditions. In fact, grass clippings are superior because they are fresh and have lost no nutrients; trees reabsorb some nutrients before shedding their leaves. In dry conditions you need to water the clippings to maintain the decomposition process. Having seen Nils's magnificent roses and impressive vegetables, I can testify that he is a true grass guru. Even rhododendrons, which will only grow in rich, acidic soil, thrive in his sandy beds. He told me he began his experiments in a greenhouse more than thirty years ago when a delivery of soil he had ordered did not turn up, forcing him to plant directly in the bed of sand he had prepared as a planting base. He covered the plants with grass clippings so they did not dry out before the soil arrived, but after seeing how fast they grew decided never to buy soil again. From then on, he planted all his plants in sand and simply covered the sand with grass clippings. I wonder if he has ever visited a rainforest. Perhaps he was an Huaorani in a previous life.

Further Readings

Åkerstedt N (2002) Boken om marktäckning och om odling i sand. Natur och Trädgård
Alm G et al (1997) Kompostboken. LTs Förlag
Forsyth A, Miyata K (2004) Tropical nature. Simon & Schuster, New York
Helldén G (1992) Grundskoleelevers förståelse av ekologiska processer. Almqvist & Wiksell International, Stockholm
Kane J (1996) Savages. Vintage Books, New York
Sjöberg F (2008) Den utbrände kronofogden som fann lyckan. Nya Doxa
Wardle DA et al (2003) Long term effects of wildfire on ecosystem properties across an island gradient. Science 300:972–975

Biological Warfare

Better a Spanish snail in the scissors
than four hundred eggs in your vegetables!

Nils Nyberg

Abstract Microorganisms produce a wealth of chemicals as they fight for space in the soil. We use some of these as medicine, such as penicillin and streptomycin, but also less well-known varieties such as cyclosporin, which revolutionised transplant surgery. Resistance to antibiotics is a growing problem in humans, but in South America there are fungi-cultivating ants which have been using antibiotics for millions of years apparently without encountering such problems. There is a lot to be learnt here. Many people see biological control as an environmentally friendly way of dealing with pests—and sometimes it works, such as when the spider mite is combated using predatory mites in tomato cultivation. At other times it is catastrophic, as when cane toads were introduced to combat beetles in sugarcane plantations in Australia. A special group of fungi makes short work of nematodes which cause diseases in both plants and animals, and a great deal of research has been conducted to get them to function as medicine or biological control agents. But profound ecological insight is required to get such a construction to work.

The soil is a crowded place and at times organisms may engage in open warfare as they jostle for space. Some microorganisms, for instance, use antibiotics to stop other species from getting too close. Sometimes we can harness the attack mechanisms of soil organisms for our own benefit.

How Soil Organisms Can Help Us

Most people have heard the story of how Alexander Fleming discovered penicillin in the 1930s by accident when a mould spore settled in a forgotten Petri dish containing yellow staphylococcus. Fleming noticed how the bacteria surrounding the mould colony died. After his death, scientists succeeded in isolating the active substance, penicillin, which has since been used to fight infections worldwide.

H. Wallander, *Soil*, DOI 10.1007/978-3-319-08458-9_6

We use many other soil organisms to produce medicines. The earthy smell we experience from rain after a long dry spell comes from special bacteria known as actinomycetes that grow in long strands reminiscent of fungal hyphae. When the ground dries out they emit spores containing geosmin, the substance that gives off the characteristically earthy aroma. It is these spores that we smell. Actinomycetes are behind many different types of antibiotic. The discovery of streptomycin revolutionised the treatment of tuberculosis, a widespread disease in Europe in the nineteenth and early twentieth century. Streptomycin is formed by members of the *Streptomyces* family, which includes various species with anti-bacterial properties.

Soil fungi have also proven to have health benefits for humans. In the 1970s, Swiss pharmaceuticals company Sandoz asked staff at its Basel office to bring soil samples back from their holidays. One of the samples, from Hardangervidda in Norway, contained hidden properties that scientists were able to develop into a lifeline drug for people with transplanted organs. When isolated from the soil

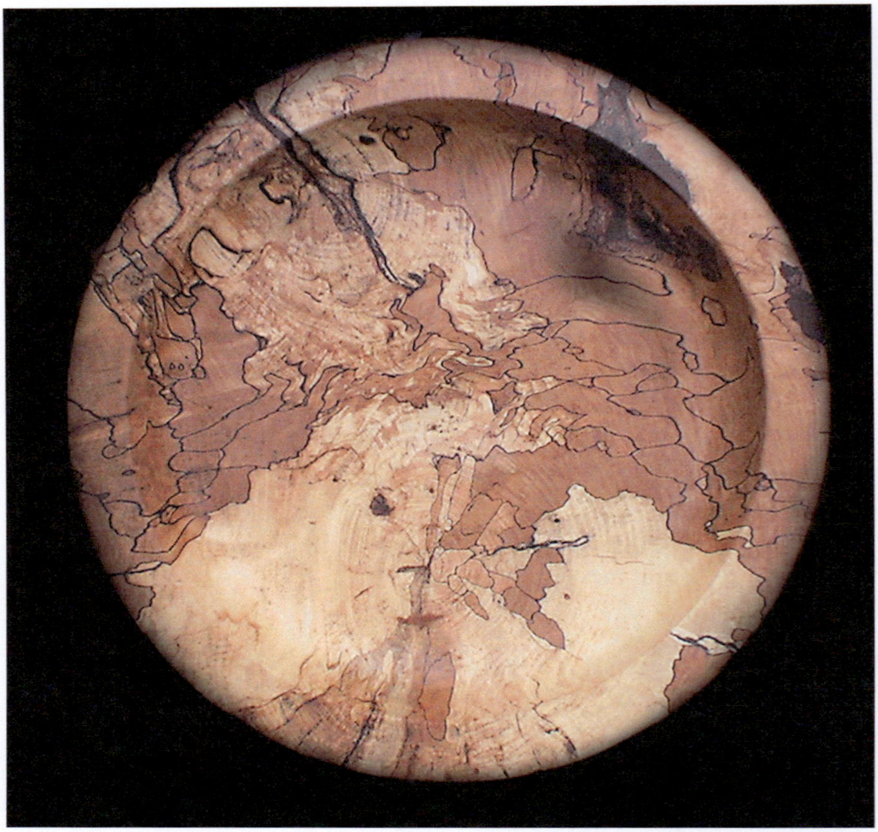

Arts of war. Species engage in a territorial struggle as soon as there is access to energy and nutrition. Capability to defend a patch is critical. When different fungi meet in a piece of wood, like the wooden tray in the picture, they secrete toxic phenolic substances that form black lines in the wood. Each individual becomes its own country, like on a map

Duped immune systems. Cordyceps, a type of fungi that infects insects, forms many active substances and has long been used in traditional Chinese medicine. For example, *Cordyceps sinensis* is thought to cure cancer. It also seems to have aphrodisiac properties and is known as the Viagra of the Himalayas. Collecting the fungal fruit bodies has become a source of subsistence for many Tibetans. When the fungi reach the Chinese market, they are worth more than their weight in gold

You need sharp eyes to spot the club-shaped fruiting body protruding from the ground, revealing the presence of an infected larva. The *Cordyceps sinensis* in the picture is related to the fungus that provides us with the drug cyclosporin, which facilitates organ transplants by inhibiting the immune system's ability to reject foreign tissue. The cyclosporin fungus *Tylopocladium inflatum* grows parasitically on beetles

sample, a fungus known as *Tolypocladium inflatum* formed a substance that proved effective at inhibiting the growth of other fungi. The effect was too limited for practical use, but luckily the scientists persisted with studying it and found that it inhibited immune response when administered to mice injected with red blood cells from a sheep. This was the discovery of cyclosporine, which stops the body from rejecting foreign tissue. It marked a breakthrough for transplant surgery and the drug is today enabling people round the globe to lead normal lives with transplanted kidneys and cystic fibrosis sufferers to survive with new lungs. All because a Swiss drugs company sent its staff on holiday with plastic sample bags.

Tolypocladium is a strange organism with an unusually high tolerance of lead. It is rare in most woodland soil but in the village of Kastad in Norway, where the ground is naturally polluted by lead, it is the commonest fungus. *Tolypocladium* is related to *Cordyceps*, a genus of fungi that attacks insects and that is also especially tolerant of heavy metals. Perhaps the pharmaceuticals industry should spend more time studying areas suffering from heavy-metal contamination in its quest for new drugs.

Magic chemicals. Plants and fungi produce all sorts of substances to defend themselves against external attack. Many have long been used in human medicine. At the Eden Project in Cornwall, modern-day shamans from Peru have painted their most important medicinal plants on the wall of an enormous greenhouse containing an entire rainforest. Tourists and school classes come from near and far to visit the Eden Project and learn about the rainforest and the people who live there. The plant in the picture is called *chiric sanango* in Spanish and grows in Amazonia

We Have Much to Learn from Fungi-Friendly Ants

Humans are not alone in harnessing the benefits of antibiotics. In a rainforest in Ecuador I watched a long line of leafcutter ants descending a tree trunk, clutching large pieces of leaf in their jaws on their way to their underground nest. Leafcutter ant nests can be several metres deep and cover tens of square metres. In the nest, ants chew the leaf pieces brought in by the workers to form a paste that serves as a substrate for a fungus that lives in symbiosis with the ants. The ants eat the tissue of the fungus, which could not survive without the care lavished on it by the ants. The ants tasked with this role are tiny and their job is to patrol the fungal colony and lick dirt from the hyphae. On their underside the ants carry *Streptomyces* bacteria which produce the antibiotic that the ants distribute in the fungal garden to keep unwanted guests at bay. It is clearly a highly successful method, considering that the ants have been cultivating the same fungus for millions of years. Over time, unwelcome microbes acquire immunity to the antibiotic, but new varieties are constantly evolving and the ants seem permanently to be one step ahead.

Antibiotic resistance is a serious and growing problem in the medical world and we humans might do well take a leaf or two out of the ants' book.

Ants can transport tonnes of leaves and are capable of causing significant damage to tropical plantations. The most attractive leaves to the ants are thin and high in sugar. When the scout ants locate a source, they emit an odorous substance and leave a trail so worker ants can find the site. Ants structure their communities on clearly defined roles. Like humans, they have colonised most environments and are particularly successful in the Tropics. In some rainforests they account for a third of the animal biomass, despite representing only a few percent of the total number of species.

I share with leafcutter ants an interest in growing fungi, and for years have dreamt of making a mushroom patch in my back garden. Mushroom growing is the perfect complement to horticulture because the dark recesses of the garden eschewed by plants make prime spots for fungi. Leftover mushrooms are also

Unconditional collaboration. Leafcutter ants carefully excise pieces of leaf of just the right size for transport to their underground nest. There they chew the leaves into a mass on which a special fungus grows and then live off the fungal mycelium. Both parties are completely dependent on this coexistence, which has been ongoing for more than 50 million years. It is remarkable that the ants have succeeded for so long without encountering the problems that have afflicted human agriculture in its short history

The leafcutter ants tend their fungi carefully. They lick the hyphae clean and remove fungal spores and other waste that can infect the colony. They also import bacteria that produce antibiotics which the ants then use to prevent undesirable bacterial growth. They place the fresh leaf mass at the top of the culture, while the white fungal mycelium grows upwards. The ants harvest the lower mycelium, which constitutes their only food

Shiitake secrets. Cultivation of the saprotrophic shiitake fungus dates back more than a thousand years in East Asia. Shiitake mushrooms can be grown on logs of deciduous trees, preferably oak or beech. The logs must be freshly cut or other fungi may get there first, making it hard for the shiitake mycelium to establish a foothold. Cultivation entails hammering wooden plugs containing fungal mycelium into pre-drilled holes

good for conditioning the soil. In the end, I decided to get the wheels moving by buying a blend of oyster and shiitake mushroom mycelia imported from the United States from an encouragingly named company called Fungi Perfecti. The mycelia arrived by post as grafts on wooden plugs that must be hammered into pre-drilled oak or beech trunks. I followed the instructions carefully and the following year was delighted to see shiitake and oyster mushrooms growing out of the trunks. That was a few years ago and the trunks seem to have lost the magic touch because mushrooms no longer grow on them. Nowadays I improve my garden soil with button mushroom compost from a local mushroom grower.

Biological Control

A continual battle for space is being played out in the soil beneath our feet, a battle in which many organisms secrete substances to impede competitors from encroaching on their territory. Some of these substances, such as penicillin and streptomycin, can be used as medicines. It is also possible to find aggressor species in the soil that can keep pests at bay in our gardens and vegetable patches. Farmers and horticulturalists have long sought to harness these biological powers rather than resort to chemical warfare, but experience has shown that it is not always so easy.

The tomato plants in my greenhouse grew well to start with. But after a few years they became thinner and thinner in spite of the household compost and chicken manure I lavished on them. On pulling up the plants I found small egg-containing cysts on the roots. The culprits were cyst nematodes—tiny microscopic worms that congregate in soil where tomatoes are grown year after year.

Crop rotation is potentially one solution to this problem. Another—if, like me, you want to grow tomatoes every year—is simply to change the soil at regular intervals. There is also a third solution, one that might save me the hassle of removing all the soil in my greenhouse. Some soils contain organisms that attack nematodes and scientists have long looked for ways to use these as biological control agents. Nematophagous fungi are carnivorous fungi that form sticky networks

Producing natural enemies. Organic pest control involves fighting pests using their natural enemies instead of toxic chemicals. Ladybirds eat large numbers of aphids and under favourable conditions can keep the aphid population in check. But biological control is also a difficult balancing act. Australia's introduction of cane toads to combat beetles that attacked sugarcane roots is just one example of pest control gone wrong. In confined spaces, as in a greenhouse, the results are usually better. Whitefly on tomatoes can be successfully controlled with parasitoid wasps and predatory mites are used to tackle spider mites

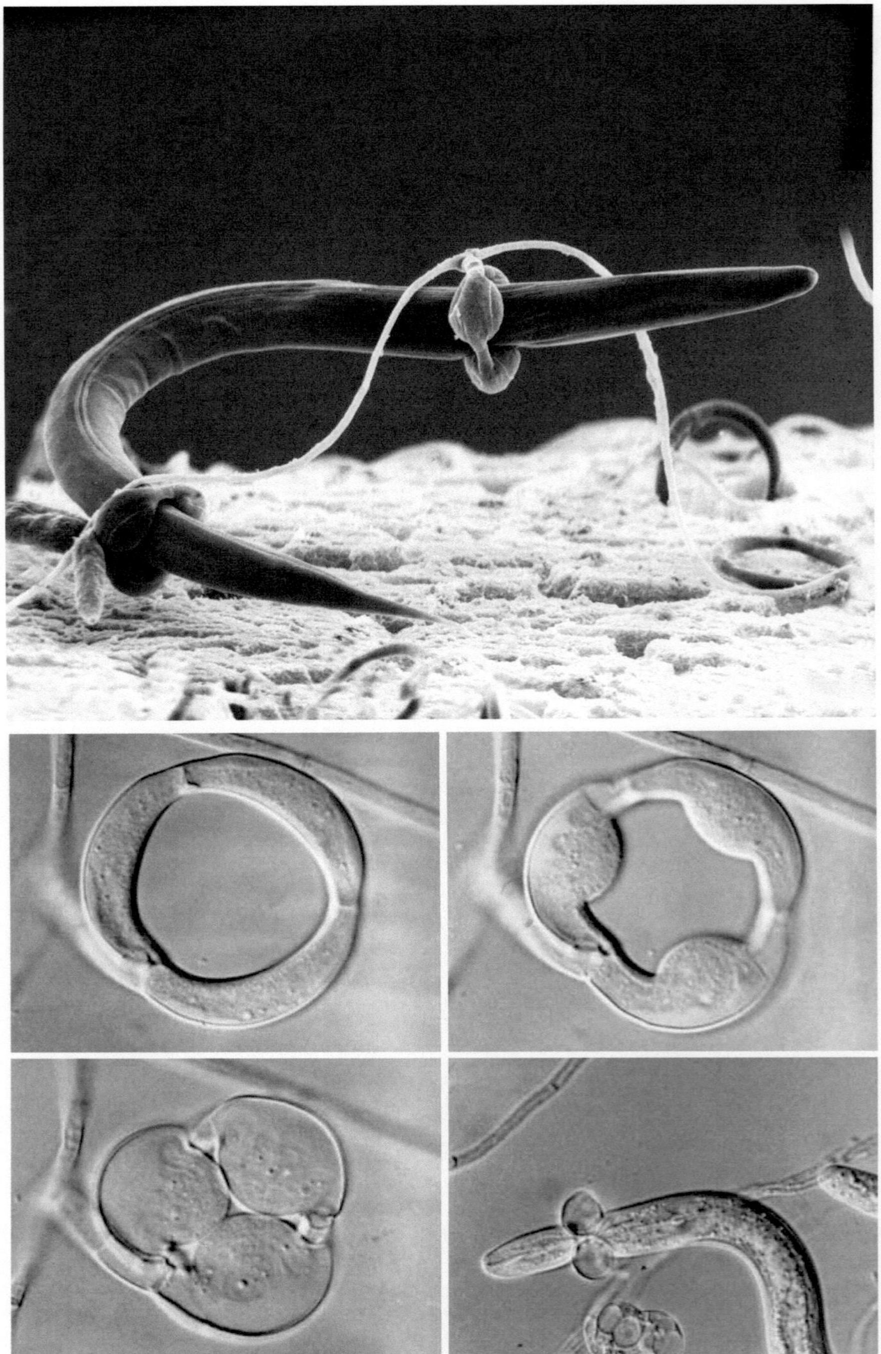

◀ *The double life of fungi*. One group of soil fungi are the fungal world's equivalent of carnivorous plants, deploying ingenious mechanisms to capture nematodes. The nematode in the photo has just swum into two retractable snares. Once captured, the fungus grows its hyphae into the nematode and breaks down its nitrogen-rich body

The image sequence shows how the snare functions. When the fungi have registered the nematode's presence, the cells swell up and trap the unsuspecting victim

or rings that trap nematodes. Once the nematode has fastened in the vice-like grip, the fungus penetrates it and fills it with digestive hyphae.

Wondering if I could enlist the help of nematophagous fungi to save myself the task of removing all my greenhouse soil, I contacted Professor Birgit Nordbring Hertz, who has spent much of her professional life studying these fungi. She explained that she had always been aware of the fungi's potential as biological agents but had chosen to focus her studies on their fascinating double life. As well as building ingenious traps for nematodes, nematophagous fungi also exist like other fungi by decomposing organic matter in the soil. Discovering why and how the fungi alternate between these two approaches has been Professor Hertz's holy grail.

The answer to this question would help me to answer my own query: is it worth investing in nematophagous fungi or is it simply better to bite the bullet and change the greenhouse soil?

It is thought that the fungi trap nematodes as nitrogen substitutes when there is a deficit of nitrogen in the soil, in the same way as sundews trap insects in and around nitrogen-poor bogs. All the chicken manure and compost in the soil means there is no shortage of nitrogen in my greenhouse, so I doubt the fungi would help.

But nematode traps do have effective uses. Ruminants like sheep and cattle are prone to nematode infestation and the roundworm, also a nematode, can infect humans. When a cow becomes infected, the nematode's eggs end up in the animal's manure, from where new nematodes hatch and lie in the grass waiting to be eaten during grazing. Once inside the cow, they develop into adulthood and lay new eggs, continuing the cycle until the cattle become ill from nematode infestation in their gut. If farmers feed nematophagous fungi to their cattle, the spores pass through the animals and end up in the manure. There they germinate and start capturing nematodes, interrupting the microorganisms' reproductive onslaught and allowing the cattle to recover. Unlike their soil-dwelling congeners, these fungi are effective at producing traps in nutrient-rich substrates like cowpats, perhaps due to the density of nematodes in this substrate.

Nematode traps have yet to catch on commercially, but proponents suggest it is just a question of ironing out a few technical issues before commercialisation can begin. The pharmaceutical companies that produce livestock drugs are less enthusiastic and would rather develop a chemical product that is easier to administer (treatments containing live fungi must be kept in certain conditions to preserve their effectiveness). However, the chemical agents used to kill nematodes are problematic, over time causing resistance in cattle and dwindling in effectiveness. Large herds are especially vulnerable to nematodes, especially if they graze for

Tame humans. Plant-parasitic nematodes can be a major nuisance for farmers, particularly in the Tropics. Many attempts have been made to use nematode-trapping fungi as biological control agents, so far with limited success. As with pet cats, which can live well on the food we put out for them and do not need to catch mice to survive, it is difficult to get the fungi to build their nematode-trapping apparatus in nitrogen-rich soil. Like cats, the fungi do as they please—and are not always bothered about pleasing humans

prolonged periods in the same area. Nematophagous fungi could be highly beneficial in such areas and also on organic farms, where the use of chemicals is banned. Fungal-based treatments will hopefully go on to play an important role in treating livestock for nematode infections in the future.

Biological warfare is a complex art. While staying in the Australian state of Queensland, I made a habit on my morning walks on the beach of studying the nightly trails left by foraging crabs. Regular patterns of tiny hollows interspersed with small piles of sand revealed the crabs' nocturnal presence. The tracks quickly wore away during the day due to human activity, only to reappear miraculously every morning. Curiosity got the better of me and I decided to grab a torch and watch the perpetrators at work. But even before reaching the beach I ran into problems: the grassy fringe along the shore was covered with toads. With every step I had to concentrate hard not to step on a toad. I am normally rather fond of amphibians, but this mass invasion was almost scary. Think Alfred Hitchcock's *The Birds*. Australians loathe the toads and some deal with them by using them as target practice for their swing. Others zigzag deliberately along roads at night to kill as many as they can.

This anti-toadism is the result of a failed attempt at biological pest control. The toads were introduced from South America in 1936 to eat beetles that were ravaging sugar cane plantations. The problem was that the toads discovered many other delicacies on the Australian menu. I think I can understand them, as I would not stick to Swedish meatballs if I were on holiday in Greece, even if they were the main dish on the hotel menu, but would choose souvlaki or octopus from the grill. How on earth can one presume that a toad will stick to eating beetles if relocated from one side of the world to the other?

In Sweden the most invasive species, at least in gardeners' eyes, is the Spanish slug. I often see a man walking along the railway line near my home carrying a bucket of salt. Every now and then he bends down and throws salt on a slug to kill it. When I last stopped for a chat his tally stood at three thousand five hundred and thirty-three slugs. First recorded in Sweden in 1975, Spanish slugs have spread rapidly. They are rumoured to eat whatever they find in their path, including other slugs, but the truth is that they merely eat dead ones. One could call them scavenger cannibals. The problem is that they also eat a wide variety of plants. Dahlias seem to be a favourite, in my garden at least. It does not take long after the first dahlia shoots emerge in late spring for Spanish slugs to appear on the scene. Once they have eaten the shoots they tunnel down to eat the stalk and roots, completely ignoring the green shoots of pale globe-thistle and *Allium* alongside. A newly sown row of salad will disappear in no time.

Perhaps we can combat the slugs with biological means. A new slug killer, Nemaslug, contains microscopic nematodes that prey on slugs and are said to kill up to ninety percent of the Spanish slugs in a garden within 4 weeks. But I doubt we will eradicate them completely and in all likelihood we will just have to put up with them. One wonders, too, what the nematodes will target if the slugs run out. Who knows, slugs may not even be the nematodes' preferred food source.

Further Readings

Hopwood DA (2007) Streptomyces in Nature and Medicine, the Antibiotic Makers. Oxford University Press, New York

Hölldobler B, Wilson EO (2009) The superorganism. WW Norton & Company Inc., New York

Nordbring Hertz B (2004) Morphogenesis in the nematode-trapping fungus Arthrobotrys oligospora. An extensive plasticity of infection structures. Mycologist 18:125–134

Sandskär B (red.) (2002) Biologisk bekämpning av skadedjur. Jordbruksverket

A Matter of Taste

The Chardonnay grape is mainly a tool for extracting flavour from the soil.

French saying

Abstract Does the soil give wine its taste? Do home-grown tomatoes taste better than bought ones because they grow in the soil instead of in a nutrient solution? Here I look for answers to these questions, in scientific literature and through conversations with tomato-growers and wine experts. Even though it is hard to find evidence for the notion that taste comes from the soil, the character of the soil affects the plants in various ways, which in turn can influence the taste. Many of the chemicals that plants create when they are exposed to the sun and to the onslaught of various parasites seem to affect taste. Limited watering of the vines improves the wine, and the best tomatoes in Tunisia come from saline soils where the plants are exposed to physiological drought.

Terroir—Is the Taste in the Soil?

Soil is not only a source of medicines and pest control but also widely considered to play a key role in determining the taste and health properties of food. I visited California one summer for the Ecological Society of America's annual congress, an event attended by more than 5,000 ecologists. During a break from the lectures and symposia, I nipped out to a local bottle shop to sample a few Napa Valley wines. I had just read *The Winemaker's Dance* by Jonathan Swinchatt and David G. Howell and was fascinated by *terroir*, the French term to describe how soil and microclimate interact with the type of grape and the winemaker's craft to give a wine its quality. The book describes the vine as a tool used to extract taste from the soil, and the authors use the special soil conditions in the Napa Valley to discuss how wines take on the character of the soil they grow in. "The combination of altitude, thin volcanic soil and water scarcity provides unparalleled flavour and intensity", read the text on the label of one of the off-licence's

H. Wallander, *Soil*, DOI 10.1007/978-3-319-08458-9_7

Hess Collection wines, from Mount Veeder in the Napa Valley. The wine was a Cabernet Sauvignon, the same grape associated with the finest Bordeaux wines. For the sake of comparison, I also chose a Petit Syrah—a grape that gives a dark-red wine—from Stags Leap, a lower-altitude Napa Valley winery. It was a Stags Leap wine that took first prize at the legendary Paris Wine Tasting of 1976, the first time an American wine beat top-quality French wines in a premier international tournament. The event marked the Napa Valley's arrival as a world-leading wine district.

Back home in Sweden, I delved deeper into the terroir concept and took part in a tasting of Alsace wines at the Grand Hotel in Lund. Here I learnt that calcareous soil results in a higher acid content, which seemed logical as I knew that many plants produce critic and oxalic acid to free up phosphorus and iron in the soil. Alsace wines are known for their mineral notes and French winemaker Jean-Michel Deiss believes the soil and climate are critical in determining a wine's taste—more critical, even, than the type of grape. At the wine tasting we were urged to filter the Alsace wine between our teeth, which the experts among us suggested felt like eating stone.

Jean-Michel plants his vines closer together than other winemakers so the roots grow deep and add quality to the wine. He believes the minerals absorbed by the roots at depths of up to 70 m below the surface give his wines their unique taste. He has many fans and his wines are regarded as being among Alsace's finest.

When you read about wine chemistry, you discover that vines exposed to stress produce a whole host of substances in their leaves and fruit. UV light damages the plant tissue, causing antioxidant pigments to form as a defence. These substances affect the taste and are believed to offer health benefits, as they also protect human tissue from different types of stress. But this is a controversial area and assorted studies have found no firm link between antioxidant intake and incidence of diseases like cancer and diabetes.

Vines also produce other substances that are said to affect human health. Resveratrol, found in grape skin, has anti-fungal and anti-bacterial properties. It has attracted considerable attention in medical research as studies indicate it enhances longevity in simpler organisms such as yeast. There are also indications that resveratrol makes energy use at the cellular level more efficient. It is well known that reduced carbohydrate intake can reduce the rate at which cells age, and resveratrol seems to have a similar effect. Laboratory tests on mice found that the animals were able to run for longer on their treadmills and put on less fat when fed a resveratrol-enhanced diet. There has also been speculation that resveratrol might explain the French paradox: that the French have a very low rate of coronary heart disease despite eating a high-fat diet.

Plants synthesise many substances to protect themselves from being eaten. Insects gradually find ways to overcome these defences, leading the plants, in turn, to develop new ones. This evolutionary game of tag produces complex plant biochemistry that winemakers can utilise to create complexity in their wines. Grape skins and seeds also contain tannins whose characteristic astringent taste is generated when they interact with proteins in the palate. As a wine matures, the tannins form

Taste of the earth? The quality of a wine depends to a large extent on vineyard location. Wine connoisseurs talk about a region's *terroir*. This French word is the basis of the quality designation *Appellation d'Origine Contrôlée* (AOC), which guarantees that a wine originates from a particular region. Growing grapes that yield good wine demands a limited water supply, soil that is not too rich in nutrients, and strong sunlight. But opinions vary as to whether the soil itself gives the wine its taste

Creative evolution. Plants are biochemical factories, their leaves full of *green* chlorophyll that captures light in the process of photosynthesis. They also contain many other pigments. The *red* anthocyanins, seen in autumn leaves after the chlorophyll has been broken down, protect the cells from the sun's ultraviolet rays. When plants come under attack from pests they produce protective substances that the invaders try to circumvent, in turn causing plants to evolve new defences in a constantly escalating arms race. The maple leaf in the photo is suffering from tar spot

longer molecules that create a softer taste because they do not penetrate the smallest mouth cavities. By attaching themselves to proteins, tannins make themselves harder to absorb for insects and other creatures that eat the plant. This is a shrewd tactic because the tannin content in grape skin is highest when the grapes have grown large in size but the seeds are yet to ripen, while tannin content drops as the grapes ripen and their sugar content rises. It is at this point that the grapes become palatable to birds, which help to disperse the seeds. Winemakers can influence the presence of taste-stimulating substances by removing leaves that overshadow the grapes, exposing them to more ultraviolet light, and by timing the harvest correctly.

Growing conditions also play a central role in how a wine develops. Excessive overwatering and soil nitrogen result in vines growing more leaves and fewer grapes. Many winemakers will tell you that it is vital to limit watering during the maturation process to ensure good quality grapes that are smaller in size with thicker, tannin-rich skins. A lower harvest volume often results in higher-quality wine, while nutrient-rich conditions and moist soil produce a higher grape volume

Noble rot. The grapes in the photo are infected with *Botrytis cinera*, a fungus that can devastate a wine harvest. But when producing wines such as Sauternes and Tokaj, the fungus is welcome and referred to as noble rot. It causes the grapes to shrivel, which concentrates the sugar content and delivers sweeter and more aromatic wines. These wines are also expensive, as the grapes must be picked by hand at exactly the right time. The most exclusive wine in this category is Château d'Yquem, which commands a price of several hundred euros per bottle

but typically watery and anonymous wine. Well-drained soil and generous sunlight yield the best results. Based on these criteria, the Hess Collection ought to be good: the thin soil of the Napa Valley offers excellent drainage and the strong sunlight gives well-pigmented grapes. Perhaps Jean Michel Deiss's Alsace wines are so good because his vines are subjected to more stress than at other vineyards. His wineries have a considerably lower production volume than the Alsace average, doubtless due to the harsher conditions on his land.

What could be better than brightening a dark autumn evening at home in Sweden with a spot of Californian sunshine? I uncorked my US wines and invited friends to a tasting with a terroir theme, adding a middling Californian blended wine to the high-quality bottles. Each guest was given three glasses of different wine and a blank sheet of paper to write notes on. The first wine was dark red with a bluish tinge. All agreed it was full-bodied and rich in tannins, with a long aftertaste: a well balanced blended wine from Robert Mondavi in California and purchased in Sweden. Next up was the expensive bottle from Stags Leap. It was dark in colour, almost inky-black, and filled the palate with a taste of dark berries. The tannins were more rounded and less astringent than the first wine. The consensus was that it was a good wine but in no way exceptional. When we moved on to the

Fruits of the Swedish soil. Wine-growing is gaining ground in Sweden and it is possible nowadays to find Swedish-produced wines in the country's alcohol stores. Several types of grape are suitable for Swedish conditions. The image shows newly harvested Nero, a sweet aromatic grape that can be both eaten and used for wine-making

last wine, the surprise in everyone's faces was instantaneous—a combined look of "What's this?" The wine exploded into the palate, its sensations crackling like an Olympic fireworks display. A wave of exultation washed through me and I heard myself saying, "Can you feel the character of the thin volcanic earth? Can you feel the intensity of the sun beating down on the Mount Veeder hillsides? Can you taste the intensity that puts the best French wines in the shade?".

My guests nodded in agreement.

How You Get Your Tomatoes to Taste Good

As a scientist, you get to experience many exotic environments on your travels. Sun-baked California contrasts starkly with Sweden's chilly lushness. Travelling helps you to see your home environment in a new light and to appreciate things you might otherwise take for granted. For me, there is nothing better than to explore the cliffs of the Kullaberg peninsula, north of Helsingborg in southern

Sweden, with the Danish coast visible across the sea. The aroma of wild garlic among the oak trees and the view over the shimmering blue sea bring back nostalgic childhood memories. On my way back home after a day spent at Kullaberg, I was driving through the region's tomato-growing heartland and decided to stop at a farm to stock up on provisions. A sea of tomatoes—red and yellow, green and deep purple, oval and pear-shaped—greeted me. I had often wondered if tomatoes must be grown in soil for maximum taste and here was an opportunity to find out: all the tomatoes here were grown in soil-free rock wool.

I asked a lady packing tomatoes in boxes if she could point out any tasty varieties.

> "They're all good," she replied. "It all depends on your personal taste."
> "But do they really taste as good when not grown in soil?" I asked.
> "Whether there's soil or not is irrelevant," she told me. "It's all about having the right fertiliser."

A young woman walked past and put down a stack of boxes with newly picked tomatoes. I could not resist asking her if she had any tips for how to grow sweet-tasting tomatoes.

Fragile taste. Soft-skinned tomatoes are often full of flavour. But because they are difficult to transport they are not cultivated commercially and are seldom found in stores. If you grow your own tomatoes, you have much more choice. Several of the varieties in the bowl come from seeds bought from Åke Truedsson, a Swede who propagates and maintains hundreds of different varieties of tomatoes, selling seeds to those who join his tomato club

> They need plenty of sunlight or they don't get sweet.
>
> But what about fertiliser, isn't that important?
>
> Not for the taste, but it's important so the plants grow well.
>
> "Who said fertiliser doesn't matter?" interjected the older woman, setting off a heated discussion. I edged diplomatically to one side and busied myself picking a suitable selection of tomatoes.

The older woman then came up and suggested I talk to Kent, the owner, whom she said knew all about the secrets of growing tasty tomatoes. She motioned towards a forklift truck that was about to disappear through a door so I sprinted after it and just managed to catch the driver's attention. Looking bemused, he asked me if anything was wrong.

> Not at all. I'd just like to ask you what's the best way to grow good tomatoes. I was talking to a lady in the shop and she told me to speak to you.

Kent grinned and took me aside to a quiet corner where he explained that it is important to expose tomato plants to water stress to ensure full flavour development. However, this is a balancing act and if you get it wrong, a lack of water will stunt the plants' growth. Also, the salt content in the water should not be too low because that causes tomatoes to taste watery. This observation reminded me of the salty soil in Tunisia that my companion Hafedh always insisted produced delicious tomatoes—a result of physiological dehydration perhaps. But the most important factor of all, Kent said, is to allow the tomatoes to ripen sufficiently while still on the plant. Some of his customers want tomatoes they can store prior to sale, which requires them to be picked before ripening. Such tomatoes never develop the right taste. Kent explained that he refuses to compromise on this point and turns such customers away.

After thanking him for sharing his knowledge, I went off to pay. At the till, the lady looked through the tomatoes I had chosen, exchanging one or two ropey-looking ones for healthier-looking specimens before saying goodbye and wishing me well with my greenhouse efforts.

On the way home I munched my way through the bag. Most of the tomatoes were nice—considerably nicer, indeed, than the ones I buy at the local supermarket. Best of the lot was an oval-shaped variety named "Santa". But did they taste as good as my home-grown ones? When I got home I went out to the greenhouse and picked a large Black Russian tomato weighing almost half a kilogram. On cutting through the thin skin, the sweet aroma of tomato wafted upwards while a well-balanced sweetness filled the palate, a sweetness with which the Rockwool tomatoes could not compete. Perhaps it was the soil or because I had forgotten to water the plants properly, or perhaps simply because Black Russian, for all its exquisite sweetness, has a fragile skin that makes it unsuitable for commercial growing.

My forays into the world of wine and tomatoes have made me realise that plant chemistry is a complex matter. Growers and winemakers can create new tastes and sensations in their produce by manipulating growing conditions. The taste does not actually come from the soil, but clearly the soil's ability to supply water and nutrition has an effect on taste and quality. Home-grown tomatoes are tastier simply because they are watered irregularly and given less nitrogen fertiliser. All the signs are that plants need a degree of stress to develop the tastiest fruit. In my view, tasteless tomatoes are grown in conditions that are too benign and because the tastiest varieties have thin skins, that render them poorly suited for commercial cultivation.

Further Readings

Goff SA, Klee HJ (2006) Plant volatile compounds: sensory cues for health and nutritional value. Science 311:815–819

Swinchatt J, Howell DG (2004) The winemaker's dance. Exploring terroir in the Napa Valley. University of California Press, Berkeley

Truedsson Å (2007) Odla tomater. Natur och Kultur, Stockholm

Soil and Climate

For someone who loves nature, who appreciates plants and animals and who finds happiness in horticulture, no problem is insoluble.

Henning E. Segerros

Abstract The soil contains more carbon than the vegetation and the atmosphere put together. How we manage our soils will therefore be of crucial significance for the climate in the future. The flow of carbon between earth and atmosphere is a priority research area; at the Abisko field station in northern Sweden, measurements are taken to determine what happens as the permafrost melts. Another major challenge we are facing is that of limited phosphorus resources. Currently, phosphorus is released via waste water, polluting lakes and seas, while the phosphorus in mines is running out. Maybe we can bind phosphorus from urine to charcoal, to make soils in the third world more fertile while reducing the proportion of carbon dioxide in the atmosphere. That is what the native populations in Amazonia did before the Europeans arrived and spread their diseases.

How Peatlands Can Fuel Climate Change

The carrots grown at Mossagården Farm, near the small town of Veberöd in southernmost Sweden, are the tastiest I have eaten: juicy, crisp, appealingly nutty in taste and organically grown. The farm stands on the site of a drained peatbog and the carrots are grown in the sandy soil left behind when the bog dried out. Drainage exposed the soil to oxygen, which together with farming led the peat to decompose and release its carbon content into the atmosphere as carbon dioxide. Just as nowadays we add carbon dioxide to the air when burning oil, so the same process took place at Mossagården, albeit more slowly. Of course, with the peat now gone, carrot growing at Mossagården is not responsible for adding to atmospheric CO_2 or fuelling climate change.

H. Wallander, *Soil*, DOI 10.1007/978-3-319-08458-9_8

Involuntary polluters. Commercial carrots are often grown in peaty soil. An unfortunate side-effect is that large quantities of carbon dioxide are released into the atmosphere as the peat decays. Carrot cultivation requires frequent earth turning, which accelerates decomposition. It is better for the environment to grow carrots in sandy soil, which emits much less CO_2. Grass and other crops that require less soil turning are good alternatives for peaty soil

But carrots are still grown on drained peatland in the central Swedish province of Närke, a practice that sees eight tonnes of carbon per hectare lost to the atmosphere every year. The process of peat degradation is accelerated by the intensive soil preparation required to grow root crops. Knowing what we do today, this type of agriculture is hard to defend. A much better way is to grow reed canary grass, which performs well in damp soil. This would obviate the need to drain the land and allow the carbon to remain fixed in the soil. It is a true win-win because reed canary grass is an excellent bioenergy crop.

In Indonesia, palm oil plantations grow on drained peatland once covered by forest. The oil is used to make "eco-friendly" diesel that we in the West can use to reduce our carbon footprint. The problem is that the carbon emissions from palm oil cultivation are so high that using this diesel is worse, from a global warming perspective, than conventional diesel. To make matters worse, peat bogs sometimes catch fire when drained, releasing even more carbon dioxide into the atmosphere. Indonesia's CO_2 emissions are higher than any other country bar the US and China. One can only hope that the powers that be devise a biodiesel ecolabel that excludes fuel produced in this way.

Large parts of Finland are also peatland. However, the Finns have been more adept at keeping the carbon in the ground, despite large-scale conversion of peaty soil to agriculture. During childhood I remember reading *Under the North Star* by the Finnish novelist Väinö Linna, a tale about a farmhand's struggles to drain a bog and build a farm. Forest grows on many of these bogs today, and though drainage accelerates peat decomposition, the presence of trees creates a balance between carbon inflow and outflow. The carbohydrates that the Finnish trees transport to their roots to facilitate nutrient uptake correspond to the carbon lost when the peat decomposes. The contrast compared to Indonesia is stark: peat degrades much faster in the Tropics and the more nutrient-rich soil results in less carbohydrate being transported below ground.

Scientists Track Carbon from Peat to Air

Most soil carbon is found in northerly latitudes, where the slower pace of decomposition allows more dead plant material to accumulate in the ground. Scientists have been studying the carbon flow between earth and air at Stordalen, a large bog near Abisko in Sweden's far north, for more than 50 years. When I visited the area in 2008 it was clear that major changes were taking place. Located in a hilly area, the bog consists of meres interspersed with palsas, which are peaty mounds with permanently frozen ice layers. If the climate warms, the permafrost melts and the mounds collapse, opening blackish cracks. When palsas collapse they are subsumed into the surrounding water, accelerating the emission of methane gas from the bog. Twenty times more potent as a greenhouse gas than carbon dioxide, methane is formed by bacteria-like Archaea in the anaerobic bog environment. As the gas percolates to the top of a palsa, it encounters aerobic bacteria that oxidise it into carbon dioxide. But if the methane instead bubbles up directly from wet areas, it goes straight into the atmosphere. Hence, any expansion of wet areas results in higher methane emissions and accelerates the pace of global warming.

Carbon Flow from Soil: The Scientists' Conundrum

Vast quantities of carbon are stored in the soil. The soil holds twice as much carbon as the global above-ground biomass. Scientists are unsure what will happen to this carbon when global temperatures rise and rainfall patterns change. Higher temperatures are likely to accelerate the decomposition of organic matter because microbiological processes work faster in warmer conditions. This would increase the amount of carbon entering the atmosphere, further raising temperatures and creating a vicious circle of climate change. Conversely, higher temperatures may lead to higher rates of forest growth and an increase in leaf and root litter, which would raise soil carbon content and bind more atmospheric carbon, mitigating global warming.

It is a complex subject. Soil consists of many different substances with myriad properties. Higher temperatures hasten the decomposition of some materials but

not others. The more complex the compounds involved, the less their decomposition seems to depend on temperature. Substances may also be slow to degrade when embedded in stable soil aggregates that decomposer enzymes cannot

◀ *Leaking soils*. Cloudberry and crowberry bushes cover palsas at Stormyren marsh near Abisko research station in northernmost Sweden. The palsas are raised peat mounds with a permanently frozen core and with grass, sedges and cotton grass growing in the wetter parts

Global warming is causing the palsas' frozen core to thaw, a process that cracks the peat (see photo), accelerates its decomposition and releases more carbon dioxide into the atmosphere. Ground methane also seeps out, fuelling the greenhouse effect. As the climate warms, increasing amounts of methane will leak from large areas of permafrost in northern Siberia

penetrate. The oxygen level in these aggregates is often low, further retarding the decomposition process. There is still a long way to go before we will be able to describe all soil components and how climate change will affect them. Models of soil carbon inflow and outflow are integral to the larger models used by scientists to project how global climate patterns will change in future. However, these models assume that decomposition of all substances in the soil will have the same temperature dependence. Researchers know that this does not reflect reality but lack the knowledge needed to make the models more accurate.

Temperature is not the only factor that affects soil carbon decomposition. Changes in rainfall can have a major impact on the flow of carbon from soil to air. Mediterranean soils, for example, are likely to become even drier in future, potentially disrupting the decomposition process there and resulting in lower carbon dioxide emissions. On the other hand, decomposition rates in damp areas may accelerate due to reduced rainfall or if marshes or bogs dry out due to higher temperatures. Lower precipitation would enable oxygen to penetrate deeper into the soil and hasten decomposition. In northern latitudes, rainfall is expected to increase—and this ought to lead to an increase in soil carbon as additional precipitation would make wet areas even wetter and less oxygenated, slowing the decomposition process.

It is clear that substantial difficulties must be overcome to improve the models for predicting global flows of soil and atmospheric carbon. Still, a number of empirical studies have been carried out in recent decades tracking changes in soil carbon. In Sweden, soil carbon volume has risen by one third since 1926, equivalent to an increase of three million tonnes per year. This corresponds to almost half of Sweden's total carbon dioxide emissions during the period. The main reason is an increase in forest biomass, which means more dead roots and leaves that eventually enter the soil and raise its carbon content. The increase has been most pronounced in the south of the country.

By contrast, Britain's soil has lost thirteen million tonnes of carbon to the atmosphere in the last 25 years, equivalent to around a tenth of the country's carbon dioxide emissions during this period and roughly the same figure as the total reduction in carbon emissions achieved by energy efficiency and other measures during the same timeframe. Carbon loss is greatest in peatlands, probably due to surface drying brought about by a warmer climate and the knock-on effect of added aeration accelerating the decomposition process.

Parched soil can lead the organisms in it to change behaviour. In acidic soil, centimetre-long white worms known as *enchytraeids* generally replace

The land that disappeared. The 1800s in Sweden saw large areas of peatland drained to increase food production for a growing population. Draining and converting peatland to arable land leads to subsidence as the peat layer shrinks and organic matter decomposes due to increased oxygen. On Bälinge marsh north of Uppsala, the land has sunk almost 2 m in places since the marsh was drained at the start of the last century. This corresponds to a couple of thousand tonnes of lost carbon per hectare. In Sweden alone, close to three million hectares of wetlands have been drained for forestry and agriculture

Growth with precision. Will forests grow faster when levels of atmospheric carbon dioxide rise? If they do, they might temporarily slow the rate of climate change by absorbing more carbon dioxide from the atmosphere. Researchers in the Alps are studying how increased atmospheric carbon dioxide is affecting trees near the *treeline*. The precision screw in the photo measures the tree's growth. Launched in 2001, the experiment has yet to record any increase in growth; experts believe another—as yet unexplained—factor is limiting the trees' growth

earthworms. Like earthworms, *enchytraeids* also dig tunnels in the soil, though they frequent the uppermost soil layer and do not mix the earth in the same way as earthworms. If you cut a cube of peat and add a few *enchytraeids* to it, the soil will look like Swiss cheese after a few days.

British scientists conducted an experiment in which they heated the soil surface to a temperature of 3° higher than normal. This resulted in the water in nearby streams turning brown because *enchytraeids*, the predominant form of animal life, tunnelled deeper underground when the hotter soil dried out. This led to increased soil aeration, accelerating the rate of decomposition and in turn leading some of the organic matter to become water-soluble and to seep with the water through the soil and into the streams. Episodes of brown water discolouration have increased both in the UK and Sweden in recent years, perhaps due to behavioural changes in *enchytraeids*. There are other potential explanations, too. Massive increases in sulphur emissions from industry in the twentieth century saw atmospheric sulphur react with raindrops to form sulphuric acid, which fell to the ground as acid rain. Acidification made the water in rivers and streams clearer because the substances that made the water murky precipitated when the pH fell. A reduction in acidification in recent decades has seen surface water regain its brownish tinge.

Important feelers. A pan-European network of masts is being built to enable more accurate measurement of the exchange of greenhouse gases between soil and atmosphere. The masts will measure national contributions to the greenhouse effect and enable scientists to compare emissions from fossil fuels with other human activities such as agriculture and transport. Precise data are required for advanced computer models that calculate carbon dioxide balances at national and continental level

Sustainable Fertiliser Use

Forests have a much higher carbon content than arable land. Growing more trees would help to sequester atmospheric carbon dioxide, though at global level this would have a negative impact on food production. There is, however, a debate on whether forests should be fertilised in order to increase tree growth and sequester more carbon. Fertiliser increases biomass production. Not only could some of this biomass be used to make renewable fuels, some of it would end up in the soil. Organic matter breaks down more slowly if fertiliser is added to the soil, which would increase the amount of soil carbon. This may sound compelling as a climate change strategy, but there are drawbacks. Walking through a fertilised young spruce forest is almost impossible. I kept getting sharp needles in my eyes when taking soil samples for one of my research projects. Fewer types of plant and fungus can survive in dense forests of this sort, creating a trade-off between positive climate impact and reduced biodiversity.

Moreover, we are unlikely to be able to afford to apply fertiliser to forests on the scale required. Producing nitrogen-based fertiliser is an energy-intensive process and a global shortage of phosphorus is also looming, as reserves in North Africa's phosphorus mines are projected to last for only 50–130 years. Phosphorus

Exemplary vehicles. We cannot continue wasting resources as we have in the past. Phosphorus is a prime example. It is a vital ingredient in agricultural fertiliser but natural deposits are dwindling and will soon be exhausted. And yet we excrete large quantities of phosphorus in our urine—only a fraction of which are recovered in water treatment plants. If we can recycle Coca-Cola cans, as in the photo, surely we can also recycle the phosphorus in our urine

prices rose seven-fold in just 14 months in 2007 and 2008 and a major headache is likely when phosphorus runs out. We must develop agricultural methods that are much less phosphorus-intensive if we are to feed the world in the future.

Constant development is driving new solutions and approaches. To reduce fossil fuel use, nitrogen fertiliser factories can run on biofuel or, as in the case of a plant currently being built in Minnesota, on wind power. This facility will produce only a fraction of the fertiliser that could be produced in an oil-powered plant, but when oil eventually runs out this may well become a preferred method. Technology exists today for extracting phosphorus and other minerals from sewage sludge, but the question is whether this will be enough. Or, to put it another way: for how many people will it be enough?

Perhaps the use of mycorrhizal fungi will catch on in the future. The fungi are disadvantaged in modern agriculture not only by artificial fertiliser but also by heavy machinery, which destroys their intricate hyphal networks. Plough-free crop growing systems are best for mycorrhizal fungi. Organic farming requires intensive mechanical intervention to keep weeds at bay, which is also deleterious to mycorrhizal fungi. On the other hand, phosphorus content is low in organically farmed soil and organic crops often have superior mycorrhizal fungi networks compared to conventional crops. Trials in Canada have shown that wheat and lentil harvest volumes can be increased by 6 and 26 % respectively if mycorrhizal fungi are added to the soil and use of phosphorus-based fertiliser is reduced by a third. Lentil yields benefit more than wheat because the latter is a grass with an intricate root system that enables the plant to fare relatively well even without fungal help. Legumes such as lentils have coarse roots that rely on fungi to absorb enough phosphorus. Estimates indicate that mycorrhizal fungi can grow hyphae whose combined length can stretch to several hundred metres in just a single gram of soil.

What mycorrhizal fungi cannot do is create new phosphorus, so we have to learn to use this resource more wisely than we do today. In Sweden, large quantities are lost in agricultural runoff into the Baltic Sea, where it causes toxic algal blooms. Farmers in south-east Asia have long used human faeces and urine as fertiliser. These are resources we will not be able to afford to ignore in the future. Only time will tell whether technological progress can enable phosphorus separation during wastewater treatment or even in our toilets. The phosphorus contained in a person's urine is sufficient to produce at least half of the food that person eats. Urine also contains nitrogen and potassium, both essential elements for plants. Perhaps we will succeed in squaring the circle to ensure enough phosphorus to guarantee our future food supply. But only if we change our habits.

Do Rainforests Hold the Key?

One of the most harmful actions from a climate perspective is to clear tropical rainforest and replace it with soybean fields or grazing land for cattle. Growing biofuels like sugar cane or palm oil is little better. Rainforests contain vast amounts of carbon that sooner or later will enter the atmosphere as carbon dioxide if the forest is felled.

Rainforest soil is poor in nutrients and has limited ability to retain nutrients like potassium, magnesium and calcium ions as they flush easily away. Another downside is the tendency of phosphorus to combine with iron and aluminium to form insoluble compounds. The rainforest flora is good at absorbing phosphorus at source when it is released by decomposing leaf litter. But once the phosphate ions seep down into the iron- and aluminium-rich soil, they quickly become unavailable.

Deep in the rainforests of eastern Ecuador I was told about *Terra preta*, the "black soil". Several million indigenous people inhabited the Amazon basin until the first white settlers arrived in the sixteenth century. The Spanish conquistador Francisco de Orellana reported large towns and well developed agriculture while exploring the Amazon River. But western diseases wiped out so many of the natives that only a fraction survived. As disease swept the jungle, communities disintegrated and were abandoned. Behind them, the indigenous inhabitants left Terra preta—one of the world's most fertile soils. The black soil is the result of the natives' mode of agriculture, a modified form of slash-and-burn in which they covered fires with earth and leaves to reduce oxygen flow. As a result, the earth was charred rather than burnt and charcoal (known as biochar in this context) was left behind.

Biochar enhances soil quality. Many people know that clay and humus-rich soils are high in nutrients and sandy soils are the opposite. Clay particles and humus have a large surface area that nutrients can attach to. By contrast, nutrients tend to run straight through sandy soil. The surface area of a teaspoon of sand grains would be roughly the size of a table tennis table, while a teaspoon of clay would cover a tennis court. Biochar is similar to clay in this respect. Magnify it and it looks sponge-like, with a large surface area for nutrients to attach to. The pores also enhance the soil's aeration. Adding biochar to a clay soil improves drainage and reduces the risk of waterlogging, while adding it to a sandy soil helps to stop it from drying out.

Through the millennia, fish and animal bones and faeces and urine from indigenous settlements accumulated in the soil along the Amazon River. Nutrients such as phosphorus, nitrogen and potassium from this human waste attached to biochar, creating a fertile, phosphorous-rich soil that was farmed for thousands of years. Normal rainforest soil cannot retain nutrients; almost all nutrients are fixed in the forest flora. When leaves and branches fall to the ground they quickly decompose and the nutrients are absorbed directly by plants. This is why rainforest clearance is so devastating: when the trees are removed, most of the nutrients disappear with them. It is usually possible to grow corn for a few years, but the soil becomes rapidly depleted and then requires 20–30 years to recover. Black soil is different: added nutrients remain locked in the soil, allowing continuing cultivation with no drop-off in yields.

Plans are afoot to add charcoal to soil to combat climate change, the idea being to sequester atmospheric carbon dioxide in biomass burned in oxygen-deprived conditions to produce charcoal. This is a simple technique that would lend itself to widespread application in developing countries as a means both to improve soil fertility and sequester carbon dioxide. It would also offer a source of much-needed income to participant nations by enabling them to join the global trade in emissions rights, which involves rich countries effectively paying poorer nations to store carbon in their soil. This enables developing countries to improve soil quality and increase food production at the same time.

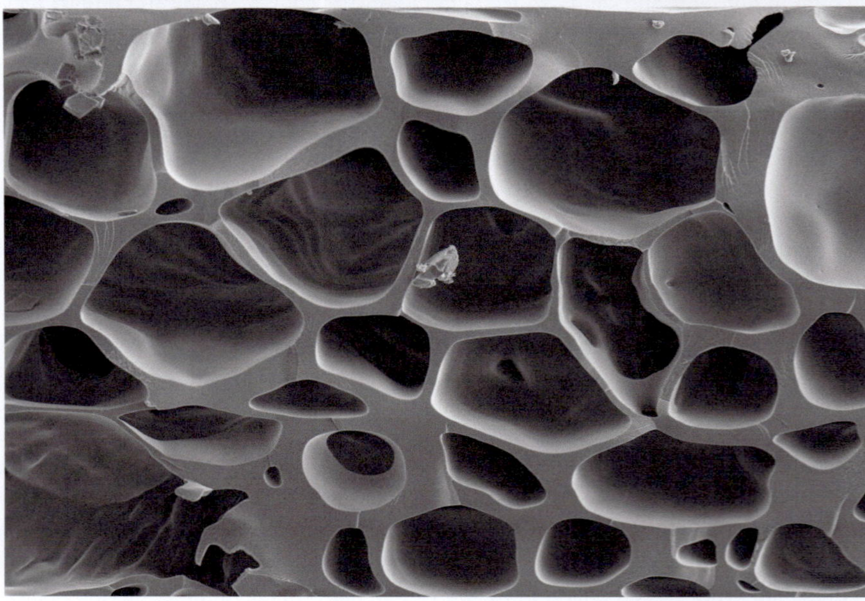

Sensitive soil. In some parts of Amazonia the soil is black and known as *terra preta*. Covering an area as large as Great Britain, terra preta was formed by a modified version of slash and burn agriculture used by the indigenous population and the gradual build-up of their food waste and excrement in the soil over thousands of years. Terra preta is highly fertile thanks to its ability to retain nutrients and water. The image shows a tract of rainforest that lacks the biochar-enriched quality of terra preta. Such soil is quickly depleted if it is planted.

Under a microscope, biochar appears hollow and sponge-like. Adding biochar greatly improves soils that are low in humus because it stores water in its cavities and nutrients stick to its surfaces. The effect is greatest in tropical soils, particularly if nutrients are added to the biochar before it is applied to the soil

Caring for the thin sliver of Earth. The highest island in the Maldives is 1.8 m above sea level and it is hardly surprising that people here want the international community to reach agreement on cutting greenhouse gas emissions. It is also important to lead by example. The Permaculture Maldives organisation produces biochar from woody biomass waste. The residue is composted or used as agricultural mulch

Various projects are being conducted to improve agricultural soil in tropical regions. At a conference in Davos, Switzerland, I met an Indian scientist involved in a project to add charcoal to arable land in southern India. The project uses special burners with reduced oxygen intake to char sticks, branches and other household biomass. The charcoal is then scattered on local fields. The initiative aroused my interest and I was put in touch with one of the coordinators, David Friese-Greene. By email he told me about banana plantations where yields had risen by 30 % after charcoal had been added to the soil during planting. He stressed it was vital to add urine or other nutrients to the charcoal before applying it to the soil. In my opinion, urine-separating dry toilets will have an important role to play in the future.

In one of his emails David asked me if I knew Folke Günther. The name was unfamiliar, but after a little research I discovered that Folke was not only Swedish and involved in biochar but that he also lived near my home. Indeed, it turned out I cycled past his house on my way to work. I sent him an email and received a reply almost immediately: he was developing a new biochar burner that he would be testing at the weekend and I was welcome to join him.

And so I joined a small group of fellow biochar enthusiasts in Folke's garden to watch the burner in action. Pyrolysis gases emitted during the charring process were put to good use when Folke lit them and used the heat to cook a wok full of meat and vegetables. Once the fire had died we found the burner full of biochar ready to be used in the garden. Folke explained that not only did the biochar fix carbon dioxide that would otherwise be released into the air, it also made his vegetables grow better. It was an enjoyable evening—and one with an optimistic atmosphere as we discussed how knowledge from the rainforest could help us look after the little sliver of Earth on which we live.

Further Readings

Belshe EF et al (2013) Tundra ecosystems observed to be CO_2 sources due to differential amplification of the carbon cycle. Ecol Lett 10:1307–1315

Cordell D et al (2009) The story of phosphorus: global food security and food for thought. Glob Environ Change-Hum Policy Dimensions 19:292–305

Gunwick NP et al (2013) A systematic review of biochar research, with a focus on its stability in situ, and its promise as a climate mitigation strategy. PLoS One 8:e75932

Novotny EH et al (2006) The "Terra Preta" phenomenon: a model for sustainable agriculture in the humid tropics. Naturwissenschaften 88:37–41

Olefeldt D et al (2012) Net carbon accumulation of a high-latitude permafrost palsa mire similar to permafrost-free peatlands. Geophys Res Lett 39:L03501

Epilogue

Soil is important to us in a wide variety of ways. In this book I have highlighted some of the aspects that interest me, both as a scientist and a layman. Overall I have sought more to explore what makes scientific research interesting than to describe the subject of soil from every angle. One of my aims was to find natural experiments in which soil formation and decomposition could be studied over long periods of time. This summer I will travel to Surtsey, an island that emerged from the depths of the Atlantic Ocean south of Iceland fifty years ago. The flora and fauna there has developed largely free from human interference, making the island a perfect place to study the role of plants and fungi in soil formation. From the moment the first plant appeared, scientists have been able to make annual visits to monitor the vegetation process and enjoy unique opportunities to study the connection between changes in the soil and the diverse array of mycorrhizal fungi that live in symbiosis with plant roots.

Among other things, the book underlines the vastness of the biodiversity found in the soil. This reservoir of organisms has provided untold benefits to Man in terms of medicines and biological agents to combat pests and parasites. But it is also important that this diversity exists for its own sake, not just ours.

Of all the topics covered, one holds special importance for our future survival: how we choose to manage the soil. I do not mean manage merely in terms of food production, but also aesthetically. How will our countryside look in the future? The sumptuous carpets of spring and summer flowers will disappear from rural areas if we over-fertilise the land. If we want a chance to curb climate change, we need to reverse the current trend of soil carbon depletion.

It is easy to feel resigned when political leaders fail to reach agreement on tackling climate change, while more and more agricultural land is destroyed by erosion and rising salinity. The future is uncertain and hard to predict, but I do see rays of hope. Many young people are pushing for a more sustainable society. Interest in permaculture and urban farming is growing. Perhaps we will unlock the secrets of the black earth of the Amazon and use that knowledge to enhance soil fertility in the developing world.

H. Wallander, *Soil*, DOI 10.1007/978-3-319-08458-9

I believe we will see a transition to more perennial crops—a transition that would decrease erosion and increase soil carbon content. This would reduce carbon dioxide in the atmosphere and slow the pace of climate change. Evidence suggests that perennial crops can be made more salt-tolerant, which would expand our access to arable land. For farmers, perennial crops offer a host of advantages. The soil can be ploughed or tilled less often, reducing diesel consumption and lessening the risk of soil compaction and erosion. Fertiliser and pesticide use would also drop, to the benefit of both farmers and the environment. Perhaps the biggest winners on all these counts would be the mycorrhizal fungi that play such a crucial role in conditioning the soil.

Proponents of permaculture have long advocated small-scale perennial crops. Trials of large-scale perennial varieties of our commonest cereal crops have produced encouraging results. When developing new perennials, we should be mindful of reducing the vulnerability of our agricultural systems and therefore base our choices on local varieties. A diverse mix of crops with high genetic variation offers greater tolerance to environmental changes and reduces susceptibility to disease. Local varieties are today disappearing at an alarming rate—a cultural bounty that is being lost. Industrial-scale growing of genetically modified crops in some parts of the world has completely replaced local alternatives and it is becoming increasingly difficult to keep these fields free from weeds and pests.

Our politicians will need to work very hard if we are to reverse the current climate trend. Short electoral terms make their job even more difficult than it already is. But that does not stop us from starting with ourselves. We do not need to worry about getting re-elected, and we can make a big difference if we make eco-friendly choices. If demand for green products increases, companies will follow and adapt their product ranges.

I visited a two thousand square metre garden in Järna, 40 km outside Stockholm, and was deeply inspired. The area represents the amount of growing space per person on Earth if we divide all arable land by the world's population. The owner's aim is to grow her entire food needs in this single plot. Some is used for animal fodder and suffices to enable her to eat meat once a week. There is no doubt we would reduce our strain on the planet by eating less meat, which is so much more resource-intensive to produce than vegetarian alternatives. The garden at Järna is an ingenious way to show how limited the resources are on our planet and how we might alter our lifestyle to live more sustainably. It is a popular destination for school trips, and arouses a great deal of curiosity and questions among children. I seemed to commit a *faux pas* when I asked the owner if she used her urine as fertiliser. Using it on her vegetables did not appeal, she told me, suggesting this could result in medical residues contaminating the soil. But the fact remains that a mix of one part urine and ten parts water provides a perfect blend of the nutrients that plants need to grow. I believe we cannot afford to waste this precious resource. In the best-case scenario, these nutrients are removed during wastewater treatment so they do not end up in our rivers and oceans. But we need to return a much higher percentage of these nutrients to the soil if we are to feed the world, especially given the fact that phosphorus supplies are running out.

We need to aim high to solve the environmental challenges that face us. For me, it is also important to find a way forward that gives inner satisfaction. I believe that many people would benefit from a deeper contact with living organisms. In *Between Earth and Sky—Our Intimate Connections to Trees* (2008) canopy biologist Nalini Nadkarni describes her research and dedication to forests. On platforms high in the treetops, she brings together people from a wide variety of backgrounds: artists and scientists, Inuit who live beyond the treeline and blind people who experience trees through other senses than most of us. New thoughts and insights emerge when these people come together to share their experiences. Feelings of kinship and respect create new bonds, perspectives and well-being. Well-being that for me perhaps reached its peak the time I scaled an olive tree near Chianni in Tuscany, where I listened to the sound of olives tumbling into the net below and watched the town of Volterra glinting like a jewel in the distance.

Think Like a Tree

Soak up the sunAffirm life's magic
Be graceful in the wind
Stand tall after a storm
Feel refreshed after it rains
Grow strong without notice
Be prepared for each season
Provide shelter to strangers
Hang tough through a cold spell
Emerge renewed at the first sign of spring
Stay deeply rooted while reaching for the sky
Be still long enough to hear your own leaves rustling.
Karen Shragg

Acknowledgments

It has been a challenge, but also extremely rewarding, to write for a broader circle of readers than the more scientific one that I am used to. A lot of people have contributed to this work and I am thankful to Lisa Christensen, Lena Ekberg, Lotta Möller, Anette Nilsson, Stefan Casta and Martin Ringh at the School of Creative Writing at Lund University for feedback and stimulating discussions. Among my scientific colleagues, I am indebted to Anders Dahlberg, Erland Bååth, Pål Axel Olsson, Edith Hammer, Birgit Nordbring-Hertz, Tryggve Persson and Ulrika Rosengren for valuable input. I would also like to thank several of my friends and family members who contributed with encouragement, comments and photographs: Helen Hasslöf, Jesper Bristulf, Erik Valeur and my wife Inger Valeur. A lot of people around the world, both scientists and laypersons, have shared their photographs for the production of this book. They are listed in the back and I am very grateful for their generosity. I am also extremely thankful to Johan Laserna for excellent work in designing the Swedish version of the book, which also forms the basis for the English edition. My editors Valeria Rinaudo and Ineke Ravesloot at Springer have been very helpful and I am very thankful to Greg McIvor and Nicole Tyszkiewicz for marvellous work with the English translation. Finally I would like to thank the Anna and Gunnar Vidfelt Foundation for Biological Research, the Royal Physiographic Society in Lund, and the Royal Swedish Academy of Agriculture and Forestry which contributed financial support for the Swedish edition of the book.

Right Holders of Photographs

Cover: Johan Laserna

Photographers with notification to page number:

Zsuzsanna Balogh-Brunstad 23, 64 (lower), 66 (upper)
George Barron 108 (upper)
Alexander Berg 92
Kerstin Berglund 128
Lynne Body 64 (upper), 102
Karina Clemmensen 97
Dolph Cooke 134 (lower)
David Chase 75
Magnus Ellström 93
Henrik de Fine Licht 105
Ari Friedland 86 (middle), 134 (upper)
Örjan Fritz 16
Paul Gardner 74
Mark Garrett 135
Reiner Giesler 29
Julie Gomez 18
Edith Hammer 10, 12, 49
Ari Jumpponen 27
June Kallestad 73
Johan Laserna 5 (right), 22, 24, 38 (upper), 48, 53,81, 82, 107, 110, 131
Jonathan Leake 5 (left)
Stefan Lindberg 60 (lower), 62 (left)
Richard Mandelbaum 106
Hanny van Megen 66 (lower)
NASA 70–71
Birgit Nordbring Hertz 108 (lower)
Pål Axel Olsson 19
Wilfred Otten 4 (upper)

Paul Perault 8
Public domain 30 (upper)
Sarah Richardson 33
Karl Ritz 66 (middle)
Rothamsted Research Ltd 2
Stefan Salmonsson 45
Michael Shane 43
Fred Stevens 95
Mikael Sörensson 50 (lower)
Suzanne Urpecz 117
Inger Valeur 58 (upper), 83, 84, 86 (upper and Valeur lower), 87, 88, 89, 90, 115, 124
Håkan Wallander 4 (lower), 6, 14, 15, 26, 30 (lower), 32, 35, 38 (lower), 44, 47,50 (upper), 58 (lower), 60 (upper), 62 (right), 98, 104, 118, 119, 126, 129, 130
Lars Wallin 76
David Walter 68
Kari Whitley 116
Daniel Winkler 103
Tomas Utsi 40
Sera Young 56
Douglas Zook 85

MIX
Papier aus verantwortungsvollen Quellen
Paper from responsible sources
FSC® C105338
www.fsc.org

If you have any concerns about our products,
you can contact us on
ProductSafety@springernature.com

In case Publisher is established outside the EU,
the EU authorized representative is:
Springer Nature Customer Service Center GmbH
Europaplatz 3, 69115 Heidelberg, Germany

Printed by Libri Plureos GmbH
in Hamburg, Germany